GEOGRAPHY AND WORLDVIEW:

A CHRISTIAN RECONNAISSANCE

Edited by

Henk Aay *and* Sander Griffioen

CALVIN CENTER SERIES
AND
UNIVERSITY PRESS OF AMERICA, INC.
LANHAM • NEW YORK • OXFORD

Copyright © 1998
University Press of America,® Inc.
4720 Boston Way
Lanham, Maryland 20706

12 Hid's Copse Rd.
Cummor Hill, Oxford OX2 9JJ

Copublished by arrangement with the
Calvin Center for Christian Scholarship

Library of Congress Cataloging-in-Publication Data

Geography and worldview : a Christian reconnaissance / edited by
Henk Aay and Sander Griffioen
p.   cm.
Includes bibliographical references and index.
1. Ecclesiastical geography. 2. Christianity—Philosophy. I. Aay,
Henk. II. Griffioen, S.
BR97.G46   1998   261.5 —DC21   98-10839 CIP

ISBN 0-7618-1042-0 (cloth: alk. ppr.)
ISBN 0-7618-1043-9 (pbk: alk. ppr.)

Contents

List of Figures

List of Figures

List of Maps

Foreword

THIS BOOK IS the product of the Calvin Center for Christian Scholarship (CCCS), which was established at Calvin College in 1976. The purpose of the CCCS is to promote creative, articulate, and rigorous Christian scholarship that addresses important, theoretical and practical issues.

The present volume is the result of Henk Aay's and Sander Griffioen's vision of bringing together an international, interdisciplinary group of scholars to discuss, within a Christian context, the relationship between geography and worldview. The scholars in the working group responded to the challenge by working very hard, not only on their own papers but in giving good critique to each other's work.

The editors and the staff of this Center are aware that the whole undertaking of looking at geography from a worldview standpoint is a discourse at a relatively beginning stage. Thus, we are pleased to contribute this reconnaissance. The book surveys aspects of the field of study and makes a kind of map, the broad contours of which can now be seen. The scholars involved in this project welcome others into this conversation, and into making this kind of cartography more exact and refined. The CCCS is delighted to present this volume for consideration by geographers and by all who are interested in the meaning of Christian thinking in intellectual life.

Grand Rapids   Ronald A. Wells
November, 1997   Director, CCCS

vii

Acknowledgments

IN AUGUST 1996, fourteen geographers and one philosopher met at Calvin College in Grand Rapids, Michigan, to consider the relationships between Christian worldview and geography. The essays in this book are based on papers read at that occasion. The conference itself was part of a larger project of the Calvin Center for Christian Scholarship, an institution within Calvin College promoting and funding Christian scholarship. Under the auspices of the Calvin Center, Henk Aay, geographer at Calvin College, and Sander Griffioen, philosopher at the Free University, Amsterdam, had previously reconnoitered the borderlands between Christian worldview and geography in Dutch scholarship and education. The conference served to provide a broad forum for their work and created an opportunity for other geographers to address this topic from different vantage points.

It is our sincere hope that this book will stimulate further scholarly interest in geography and worldview, in general, and promote the development of a Christian mind in geography, in particular. We express our profound gratitude to the Calvin Center for Christian Scholarship, to Calvin College, and, in particular, to Ronald A. Wells, director of the Center, as well as Donna Romanowski and Amy Bergsma, administrative staff of the CCCS, for providing pivotal support and principled direction for this scholarship, for the conference, and for this publication. Last, and not least, we thank the Department of Philosophy at the Free University, Amsterdam, for facilities and support that made it possible for the editors to work together in the same place during the final editing of the manuscript.

Henk Aay
Sander Griffioen
Easter, 1997

Introduction: Two Approaches

TWO AVENUES ARE open to the pursuit of Christian scholarship. First, it is possible to start outside the special disciplines and ask what consequences basic beliefs or fundamental notions may or should have for scientific research. What one is after in this case are the implications *for* the disciplines. The other approach takes as its point of departure the discipline itself. It starts, as we say, where the action is, in the daily praxis of scientists and with the results of their research, and then sets out to explore, elucidate, and extrapolate the implications of their work in order to identify the implicit beliefs about the world. Here we are dealing with implications *of* the practice of science. In this case, the trajectory followed is not from the wider context of beliefs, convictions, and general philosophical notions toward the everyday practice of the special disciplines but is rather the other way around, starting at the work floor and moving through the work of interpretation and extrapolation toward wider horizons.

It is important not to view the two movements here distinguished as mutually exclusive. Christian scholarship can only develop by working both from the inside and the outside of a discipline. Put differently, Christian scholarship needs normative as well as implicit approaches. It needs bold visions directly fed by root convictions, and it needs the careful work of interpreting and questioning the practice and products of science as well. We believe the two should be connected. On the one hand, the relevance of those broad visions needs to be shown in terms of concrete scientific work. On the other hand, analyses and interpretations need a vision as a compass. If, in general, a healthy balance is maintained between the normative and the implicit approaches, the question of whether to take one or the other in a specific case will depend on circumstances. Of course, it is also a matter of the division of labor within the community of Christian scholars: Some are better than others in asking the big questions—here too one finds foxes and hedgehogs (Berlin 1957). However, tensions can develop quite easily. If the normative approach dominates, then an unfruitful apriorism will result. In relation to the worldview theme, such an apriorism may take the shape of a "religious totalism," traces of which Nicholas Wolterstorff (1989) detected in the tradition of Calvin College and the Free University (see the concluding chapter in this book). Such a totalism becomes evident, for instance, when it is held that scientific results are entirely determined by one's worldview. The pendulum is more often allowed to swing far to the other side within the scholarly communities. Then Christian scholarship becomes a matter of whatever these scholars happen to do, without a (search for) communal direction, without a worldview to give orientation. Then the only question such scholars ask is whether the implicit beliefs of the reigning scientific paradigm are consonant with the Christian faith.

The discussion thus far has an immediate bearing on the worldview theme. In fact, a major reason why we have decided to focus this study on worldview rather than on other frequently used concepts such as paradigm or discourse is that the former can serve quite well as a vehicle for both approaches outlined above, the implicit as well as the normative, whereas the other two cannot be used well in a normative sense. For an example of this let us turn briefly to a more or less arbitrarily chosen definition of worldview, Clifford Geertz's description of the communal worldview of a people: "Their world view is their picture of the way things in sheer actuality are, their concept of nature, of self, of society. It contains their most comprehensive ideas of order" (Geertz 1973, 127). Obviously, here an anthropologist is at work defining a worldview of a native people rather than of a scholarly community. Nevertheless, it is not too far fetched to adapt his definition to our purpose. A worldview then can be said to provide a comprehensive scheme of the surrounding reality. What is significant is that such a definition can be used in both ways, normatively and implicitly. The former applies when a worldview is introduced as expressing core beliefs about the world, about human destiny, about what makes for sense and non-sense, and so forth. Such a worldview then can be made to function as an interpretative framework for all kinds of fields of human action, including scientific activities (Wolters 1991, 239). In the latter sense, however, worldview is a container concept for beliefs that are implicit to certain traditions, paradigms, or schools of thought. An influential theoretical account of the implicit worldview is offered by Peter Berger and Thomas Luckmann's (1966), *The Social Construction of Reality*. It only treats worldviews as factually present comprehensive schemes of interpretation, the social function of which is to provide stability to a particular symbolical order. Worldviews then belong to the large class of social defense mechanisms, having no (or any marginal) explorative, disclosing functions. Characteristically, these authors remain silent on the question of whether their own sociology of knowledge depends on some worldview.

However, our argument thus far is not meant to discard the implicit approach. The ideal picture would be one of a cycle with the implicit and normative approaches as distinct but connected phases. The worldview then would not only function as a *terminus a quo*, the point from where we start, the compass used to find our orientation in the world of scholarship, but also as a *terminus ad quem*, a point to which we return, and which itself then is transformed and enriched by the results of this scientific work. It is important that the impression be avoided that a Christian worldview is a fully developed set of principles ready to be applied to any field of science. Such an idea would open the door to religious totalism. A full-fledged worldview not only harbors general principles directly related to basic convictions—Wolters (1985) offers a valuable survey of biblical principles—but also reflects experience about how the world is structured. In practice, however, it is far from easy to formulate such a fully developed worldview. Therefore, it is only realistic to anticipate that an academic forum is likely to judge the fruits of Christian scholarship as either too aprioristic or too implicit.

This makes it clear that to form a Christian mind in geography will never be simply a matter of lockstep application of ready-made principles to the field. It can only be done as part of a communal scholarly quest. The chapters of this book in one way or another follow the two strands distinguished here. On the one hand, we see human geographers busy tracing developments within the discipline (or broader within the social sciences) to identify spots where Christian scholarship would make a difference. On the other hand, the authors look at their discipline from a certain distance to see what implications broader developments have for geography.

To some extent all chapters follow both approaches in one way or another because the authors want to speak as geographers also when addressing wider cultural and religious issues with a view to the implications for their own work. Nevertheless, the chapters are grouped in such a way as to show a distinction between the two approaches. In the first four chapters, the emphasis is on the movement from inside to outside: these all take their point of departure in the praxis (past or present) of geographers, and from there on show the broader implications. David Livingstone lays bare the persistent relevance of assumptions that were once derived from natural theology. These assumptions were ostensibly ousted by positivism's pursuit of scientific neutrality. The succeeding chapters, written by David Ley, Iain Wallace, and Janel Curry-Roper, each in its own way analyzes the present breakdown of positivism. To grasp the full consequences of this collapse, it is necessary, these geographers believe, to understand that positivism (modernism and so forth) only seemingly functioned as a neutral methodology but, in fact, had all the qualities of a specific worldview and as such was instituted as a paradigm in geography. Understood as a worldview, the social function of positivism was, as Berger and Luckmann would put it, to stabilize and monopolize a specific (and by no means self-evident) interpretation of reality. This meaning of worldview becomes especially transparent in Iain Wallace's chapter. The breakdown of the positivist paradigm could not but unleash forces of diversity and dissent.

The one overriding question asked in these chapters is what the present postmodern spirit offers. As the norm of scientific neutrality is no longer de rigueur, not even in mainstream science, what does this mean for Christian scholarship? Interestingly enough the emphasis then shifts to implications for geography! One benefit of the present tolerance of worldview diversity is that it marks the end of a once dominant neutrality ideal that left no place for expressions of alternate convictions. As David Livingstone insists: "Christian geographers . . . are entirely within their epistemic rights to pursue projects in keeping with their core beliefs" (Livingstone, below, Ch. 1; also Harrison and Livingstone 1980). Similarly Curry-Roper toward the end of her chapter explores models of covenantal relations as an alternate to positivism. In part, this answers the central concern of David Ley's chapter: how to prevent the tolerance toward diversity from turning into a new relativism. The solution he seeks is a relationalism that escapes the dilemma of boundless diversity or rigorous universalism.

The approach from outside to inside geography finds ample elaboration in the second group of chapters, starting with Gerda Hoekveld-Meijer's chapter on the significance of externality fields and borders for a Christian geography, followed by Gerard Hoekveld's spelling out the implications for geography of a biblical understanding of citizenship, and Henk Aay's study of the impact of neo-Calvinism on geographic education in the Netherlands.

Gerda Hoekveld-Meijer brings newly gained knowledge from a second Ph.D. (in theology) to the foundations of geography. The biblical themes of regeneration, justice, and encounter are given a geographical meaning in such core concepts as regional inequality, externality field, and border. These should become geography's principal working concepts and reform its academic mission. In a somewhat parallel chapter, Gerard Hoekveld argues that the concepts of citizenship and externalities provide a normative orientation for geography—one that distinguishes righteous and unrighteous regional differentiation.

Henk Aay's contribution makes clear that a worldview program involves promises as well as risks: promise because neo-Calvinism could indeed reform geography and risk because the effort lacked a connection between what we have distinguished as implications for and implications of. Looking at the history of the Calvinist school movement in the Netherlands, Aay concludes that it was a long road from a worldview to geographical textbooks. The tale he tells is partly one of roads not taken, of a worldview not growing into a tradition with a certain paradigmatic significance. Finally, Sander Griffioen's chapter takes the discussion to a meta-level: It bends back to the worldview notion itself to examine its benefits and hazards for the social sciences. It also attempts to assess the strength of the tradition shared by the Free University and Calvin College—a tradition that owes much to Abraham Kuyper whose thought with some justification has been characterized as "postmodern before the time," a tradition, moreover, that retains its vitality (Hart 1984; Clouser 1991; Griffioen and Balk 1995).

Henk Aay
Sander Griffioen

Chapter 1

Geography and the Natural Theology Imperative

David N. Livingstone

> The object of this address is, first, to direct attention to the enormous service of science in liberating our minds from their century-long subjection to ancient dogmas. . . . [A]ll religions, ancient and modern, including Christianity, are, like all the sciences, wholly of human origin . . . . By reason of a strange concatenation of ancient events, the theological doctrines that have so long dominated European Christianity are the outgrowth of a series of crude, superstitious beliefs, which originated several thousand years ago among the barbarous, ignorant, credulous peoples of southwestern Asia; a body of beliefs which was recorded, in the form commonly known to us, by a people who believed themselves 'chosen' by their god from among all other peoples; a body of beliefs was introduced into Europe in close association with a new gospel which that 'chosen' people rejected, a rejection which later caused three centuries of cruel persecution. And yet we—that is, all peoples of European stock—adopted both the old beliefs and the new gospel as the infallible 'Word of God.' Does history record anything more extraordinary?

THUS BEGAN THE 1933 Hector Maiben Lecture delivered to the American Association for the Advancement of Science held in Cambridge, Massachusetts. With such sentiments, it is hardly surprising that the speaker should laud the two "masterly volumes" by Andrew Dickson White on *The Warfare of Science and Theology in Christendom* and recommend to his hearers that those "who have not read this great work should do so without delay." Why? Because they revealed how "Christianity . . . arrested the normal development of the physical sciences for over fifteen hundred years" (quoted in Chorley, et al. 1973, 759-63).

Three days after Christmas 1933, these words were uttered by the eighty-three year old William Morris Davis. It was to be his valedictory speech, for within a few weeks, Davis was to suffer a heart attack on the first of February 1934 and die within four days. A brief memorial service would be held at the Neighborhood Church, a nonsectarian institution apparently advocating a kind of scientific naturalism, where Davis himself had spoken on one or two occasions (see Chorley et al. 1973, 728-29). The appropriateness of the service was assured as the minister, Dr. Soares, thoroughly acquainted himself with the scientistic faith that Davis had so recently voiced under the hugely appropriate title, "The Faith of Reverent Science." Davis articulated, in extenso, his scientific credo. Positivistic in tenor, evolutionary in spirit, and rationalistic in temper,

Davis de-supernaturalised revelation, mythologised the Fall, Darwinized the history of religion, and demonised those "very orthodox parents" who "taught their children an awful catechism" (Chorley 1973, 767). It all added up to a scientific faith. And because scientific rationalists had not "organized themselves into a religious body" nor "formulated a creed," he himself declared their principle "articles of faith," faith in intellectual progress, in "the persistence of order," and in self-sacrifice for the common good (Chorley 1973, 775, 781, 776). If the professorhood could triumph over the priesthood and enlist in its services those who recognized "the victory of modern science over ancient theology," then society had a rosy future. If the "study of theological apologetics" were to be replaced "by a scientific study of the nature of modern man," the festival of a scientific eschaton could be celebrated. Future historians would "look back on this period as one in which many have liberated themselves from the centuries-long enslavement to the crude myths of an ancient and ignorant Asiatic people. The historians will see, as we also may see, in the long period during which the enslavement lasted, a measure of the astounding credulity of the human mind and of its incapacity to think out its problems by a reasonable, scientific method" (Chorley 1973, 784-85). To sum up: Davis's so-called "reconciliation" of science and theology was, in reality, a subversion of the distinctive doctrinal elements of the Christian religion in the interests of scientific chauvinism.

I use these reflections on W. M. Davis as the occasion to illustrate something of how the kind of antimetaphysical stance here voiced has deeply conditioned the intellectual management of the geographical enterprise. First, and at much the greatest length, I suggest that the triumphalist image of religion's defeat at the hands of scientific rationality has encouraged the production of skewed histories of the tradition in which the theological has been expunged in the interests of secular modernity. I contend that we have been served an impoverished historical menu and I shall try to illustrate this by uncovering some of the intimate connections, whether for good or ill, between geographical inquiry and the tradition of natural theology. Second, I hope to be able to illustrate some of the ways in which this corpus of work was connected up with foundationalist epistemology, to tease out some of the implications of that association, and to hint at some strategies for coping with foundationalism's increasing dissolution at this contemporary postmodern moment. Third, the recovery that I will try to effect here should—if all goes according to plan—point up some key historiographical issues that I believe ought to feature more prominently on the subject's current agenda. The silences and absences so fervently highlighted by feminists and postcolonialists, for example, do not, in my view, exhaust the major elisions in the recounting of modern geography's narrative. And finally, I shall want to suggest some of the ways in which those with Christian commitments have been too captive to the fashionable tropes of the new academic establishment and venture one or two observations, drawing on the writings of philosophical colleagues, on taking courage to pursue different agendas.

The Natural Theology Imperative

The implications of the kind of conflict historiography promulgated by Andrew Dickson White (see Russell 1989) and ardently endorsed by W. M. Davis, were indeed far reaching. Within geography, this was already detected, as far back as 1951, by George Tatham who, in surveying the evolution of the discipline in the nineteenth century complained about the way in which Ritter's teleological perspective had been treated by historians "as though it impaired the quality of Ritter's work, making it somehow unscientific. There is, of course no basis for such criticism," he went on, "Contemporary developments in science have shown that a teleological philosophy can be combined with the most rigid scientific accuracy in research and there is every indication that in Ritter it was so combined" (Tatham 1951, 47-48).

Since then the warfare model of the history of science and religion has been dismantled with forensic precision by a variety of historians of science (most recently Brooke, 1991). This does not mean to imply, of course, that there have never been conflicts between scientists and theologians, or squabbles over cultural authority between a new professionalizing science and the old clerical guardians of Scripture and social status. What it does indicate is that the history of Christianity's engagement with science has not been characterized by systemic strife. The inadequacy of such pugilistic dramaturgy within geography could be illustrated in many arenas, and I will hint at some of these in due course. But for the moment, I want to attend to the tradition of natural theology and illustrate something of its profound impact on geography's history (see D. N. Livingstone 1984, in press; Glacken 1967; Tuan 1968).

The project of natural theology—in contrast to revealed theology—has conventionally engaged in the attempt to derive certain claims about God without recourse to scriptural revelation. Its most characteristic formula, perhaps, has been the design argument that postulates that the existence of a superintending divine intelligence and certain of His attributes can rationally be inferred from evidences of sagacious planning in the world. The tradition's classical formulations, such as those of Augustine, Aquinas, and Scotus predate the scientific revolution; but during the seventeenth century, empirical investigation of nature was drawn more centrally into the orbit of teleological confession. A major stimulus was the Boyle Lectures—a series of sermons provided for in Boyle's will to vindicate Christianity against infidelity, and from the outset, Newtonian science was deployed to serve this end.

Scientific natural theology could meet a variety of needs (Brooke 1989; Brooke 1991, 192-225). It could help obviate the seeming conflict between believing piety and the infidelity of mechanical theories by projecting images of a divine mechanic or architect; it could facilitate the presentation of potentially seditious science in religious vocabulary thereby shielding scientific practitioners from suspicions of heterodoxy; and it could provide answers to specific problems, such as postulating creationist explanations for patterns of biogeographical distribution. Natural theology had political and social uses as well. In the seventeenth century, for example, analogies between the laws of matter and the role of monarchy enabled particular political arrangements to be underwritten by both natural and divine law (Jacob 1976; Shapin 1981). Such conceptual elasticity reveals substantial variation in the natural theology tradition—a variation that itself has a geography. In England, Germany, the Netherlands, and America, physico-theology survived much better than in France where connections between Catholic orthodoxy and dissent were more polarized.

Geography's engagement with natural theology was wide ranging and encompassed a broad variety of themes: Divine design was detected in the earth's surface features, its plant and animal life, its demographic characteristics, its regional articulations (see Glacken, 1967). Such topics were centrally present in the classic seventeenth- and eighteenth-century statements of physico-theology and persisted within geographical discourse throughout the nineteenth and even into the twentieth century. In the *Wisdom of God Manifested in the Works of Creation* (1691), for example, John Ray called on the hydrological relations of the globe—the migrating instinct of birds, the human anatomy, and vegetation—to show how form and function expressed an overall harmony that attested to divine beneficence. By contrast, it was the *im*perfection of the earth's features that, to Thomas Burnet, revealed its fittedness for sinful humanity. In his *Sacred Theory of the Earth*, he argued that throughout its history the planet had decayed from original perfection, a dissolution in which the deluge played a key role. This gloomy rehearsal of the global story, however, encouraged others to present a more flattering picture of the contemporary earth and its suitability for humankind. Thus John Woodward, in his *Essay Towards a Natural History of the Earth* (1695), could see little evidence that the pre- and postdiluvial worlds were radically different. For him, as Gordon Davies (1969, 116) puts it, the "world which emerged from the diluvial metamorphosis was a world perfectly adapted to the needs of fallen man." To Woodward, the close links between the accumulation of humus, soil erosion, and human agriculture tellingly revealed divine design.

Perhaps the central eighteenth-century work in this genre was William Derham's *Physico-Theology* of 1713, which was devoted to an exposition of how the "Terraqueous Globe" provided "A Demonstration of the Being and Attributes of God from His Works of Creation." The detail of Derham's volume repeated many of the standard themes already rehearsed by Ray but his application of the design argument to population theory—a move about which Süssmilch enthused—constituted a significant development of the tradition. Certainly others had incorporated population statistics into the fabric of natural theology. John Graunt, for example, elaborated in his *Natural and Political Observations Made Upon the Bills of Mortality* (1662) on how Christianity and the laws of nature alike endorsed the moral wisdom of monogamy, while William Petty connected up population growth and distribution with physico-theology. But Derham elaborated on the way in which demography disclosed how the "One that ruleth the World" had kept a balance of population:

> Is it possible that every Species of Animals should so evenly be preserved, proportionate to the Occasions of the World? That they should be so well balanced in all Ages and Places, without the Help of Almighty Wisdom and Power? How is it possible by the bare Rules, and blind Acts of Nature, that there should be any tolerable Proportion; for Instance, between Males and Females, either of Mankind, or of any other Creature; especially such as are of a ferine, not of a domestic Nature, and consequently out of the Command and Management of Man? (Derham 1727, 180-81).

Derham's pioneering observations on sex ratios, demographic balance, and population structure were thus domiciled in the framework of teleological confession. And it is precisely this kind of advance that defies the categories of those historians' pursuing a ubiquitous conflict between science and theology. This theme of the balance of nature, now regarded as so central to the genesis of ecological thinking, also played a prominent role in the writings of those like Gilbert White who wrote of the economy of nature. In *The Natural History of Selborne*, published in 1789, White recorded the natural order of his own parish, surveying everything from taxonomy and ornithology to seasonal change (Worster 1977). The area's complex unity in diversity attested to the God who had made "nature . . . a great economist." Here was a political economy of nature; everything fitted together "economically" (see Livingstone, 1995). Similar sentiments are clearly discernible in the writings of Linnaeus, K. L. Willdenow, and Eberhard Zimmerman (see Larson 1986).

Despite a range of philosophical assaults during the Enlightenment, emanating from such thinkers as Kant and Hume, the use of geographical data in natural theological argument survived well into the nineteenth century and beyond. Much of it took as its point of departure William Paley's extraordinarily popular *Natural Theology* of 1802, a volume that Darwin had found inspirational in his early days. Geography's complicity in the natural-theology enterprise in this period can adequately be illustrated by reference to two discrete bodies of literature, namely, Christian apologetics and geographical science.

The use of geographical data in works of popular Christian apologetics has been sadly neglected by historians of the subject. Here I can only hint at its significance by reference to two works of the genre from either end of the nineteenth century. In his book *The Christian Philosopher or the Connection of Science and Philosophy with Religion,* first published in 1823, Thomas Dick (1774-1857) devoted considerable attention to geography as one of the "Sciences which are related to Religion and Christian theology." Dick, a liberal Scottish thinker, impatient with theological technicalities, developed a natural theology that enjoyed a particularly large readership in the United States (Macpherson 1951). Here he fastened on such topics as the figure of the earth, the natural and artificial divisions of the globe, the features of mountains, oceans, and rivers, and population size as indicative of the operations of divine providence. Such sentiments were, by now, conventional enough. But geography was of even more specific interest to religious believers due to its intimate connection with the missionary enterprise. As Dick put it: "In a religious point of view, Geography is a science of peculiar interest. For 'the salvation of God,' which Christianity unfolds, is destined to be proclaimed in every land. . . . But, without exploring every region of the earth . . . we can never carry into effect the purpose of God." Accordingly, directors of missionary enterprises were advised to acquaint themselves with geographical knowledge. Christianity therefore had nothing but the most intimate interest in contemporary "voyages of discovery" because they were engaged in bringing to light the "moral and political movements which are presently agitating the nations" (Dick 1825, 237, 239). This ethical diagnosis was particularly significant, for alongside its topographical disclosures, geography was engaged in nothing less than providing a moral inventory of the globe. In this way, geography played its role in the cultivation of a moral teleology. Geographical knowledge was thus presented as the stimulus par excellence to embarking on a global moral crusade to bring benevolence and enlightenment to the ends of the earth. Thus, if geography could confirm faith through its elucidation of providential arrangements of the earth's surface, it could also stimulate the faithful to remake the earth through worldwide evangelization. Given these sentiments, it is entirely understandable that David Livingstone (1857, 4) was gratified to find in Dick's work confirmation of his "own conviction that religion and science were friendly to each other, fully proved and enforced."

My second example comes from a book described on its title page as "especially suitable for Sundays both in schools and private families." It was authored by the Rev. Dr. Brewer of Trinity Hall Cambridge, and was entitled *Theology in Science; or, The Testimony of Science to the Wisdom and Goodness of God.* A variety of sciences including geology, geography, and ethnology were surveyed and their findings presented in catechetical form. Thus, divine wisdom and goodness were disclosed in such topics as continental configuration and orientation, the environmental modification of the human race, the reciprocal dependence of the Old and New Worlds, the geography of religions and the physical geography of the sea. The underlying assumption was that the "Great Being, who 'spake and it was done,' commanded what was made to be made in its present fashion, and not in some other" (Brewer n.d. [circa 1875], preface). When we record that the volume included regional recitations by continent, the educational significance of documents such as these is particularly striking. Indeed, we should recall that a century and more earlier Isaac Watts (1736) had written a geography text for students and John Wesley recommended the study of geography, after grammar and arithmetic, in the pursuance of self-education (see Jeffrey 1987, 244-45).

Geography thus had the capacity to serve the pedagogic sentiments of Victorian Christian apologists. It could also be used by more serious philosopher-theologians. For example, in his Baird Lectures for 1876-77 (Flint

1879, prefatory note), the Edinburgh Professor of Divinity, Robert Flint, "the most unjustly forgotten philosopher of the nineteenth century" according to Alasdair MacIntyre (1990, 22), was still turning—albeit with much greater philosophical sophistication—to the writings of such geographers as Ritter, Kapp, Duval, and Guyot for what he called their "rich store of teleological data as to the fitness of the earth to be the dwelling-place and the schoolhouse of human beings." Particular praise was reserved for Maury's work on the physical geography of the sea. Thus, Flint appended to his first lecture series a section on "Geology, Geography, Etc. and the Design Argument." But he took pains to insist that advocates of geographical teleology must not cave in to the fashionable notion of the relations between nature and culture "as an immediate and inexplicable pre-established harmony like that which Leibnitz supposed to exist between the body and soul." To the contrary, they should be interpreted in terms of "cause and effect, or action and reaction, or mutual influence" (Flint 1883, 376). To support this claim, he called as witness the French philosopher Victor Cousin, an enthusiastic advocate of the Scottish Common Sense School of philosophy (see Davie 1986, 70-109), who in his *Philosophy of History in France and Germany* spelled out the role of geography in history in such a way as to subvert those who too tightly welded together people and place. As Cousin had already put it:

> The belief in a pre-established harmony between man and nature is, indeed, considerably more absurd than in a pre-established harmony between the body and soul; . . . every country is not created with a people in it, nor is every people permanently fixed to a particular country. Imagination may be deceived for a moment by an obvious process of association into this belief of certain peoples being suited for certain lands, independently of the action of natural causes . . . Besides, just as the dictum, 'Marriages are made in heaven,' is seriously discredited by the great number that are badly made, so the kindred opinion that every country gets the people which suits it, and every people the country, as a direct and immediate consequence of their pre-established harmony, is equally discredited by the prevalence of ill-assorted unions, a great many worthless peoples living in magnificent lands, while far better peoples have much worse ones (quoted in Flint 1883, 377).

That such sources could be drawn upon by writers of popular apologetics and philosophical theology bears witness to the continued vitality of teleological thinking within geography itself. This indeed was recognized by H. R. Mill who, writing in 1929 on the development of the subject in the nineteenth century, commented that "teleology or the argument from design. . . . was tacitly accepted or explicitly avowed by almost every writer on the theory of geography, and Carl Ritter distinctly recognized and adopted it as the unifying principle of his system." In his Bridgewater Treatise on *Astronomy and General Physics* (1834), for example, William Whewell used the phytogeographical work of Humboldt to demonstrate how the Creator had adapted various plant forms to their regional climatic regimes. This, by the way, was a common strategy among English reviewers of Humboldt who, disturbed by the absence of theological confession in his writings, made every effort to teleologise the Humboldtian Kosmos (Rupke in press, Introduction). Again, Mary Somerville, in her *Physical Geography* (1858), argued that the patterns of human settlement demonstrated the arrangement of divine wisdom, while Arnold Guyot's ecological geography, drawing inspiration from Ritter, was built on the providentially governed "grand harmonies" of nature. Thus Guyot (1897, 53) concluded his investigation of continental relief with the comment that "all is done with order and measure, and according to a plan which we shall have a right to believe was foreseen and intended." That this was a driving force throughout Guyot's life is plain from his strenuous and immensely influential efforts to provide a concordist reading of Genesis and geology (Guyot, 1884) and in his cultivation of his protégé, the paleontologist William Berryman Scott, the grandson of Charles Hodge, to whom he wrote just after his appointment to the Princeton faculty: "Your mind is not shallow enough to allow you, while admiring Nature, to seek in it what God never put in it, & to believe, as too many do, that true Science is possible when we confine our view to the narrow field of this lower domain" (cited in Gundlach 1995, 187). As for Matthew Fontaine Maury, author of the *Physical Geography of the Sea* (1855), it was the mechanistic operations of marine and atmospheric circulation systems and of energy transfers between land, sea, and air that confirmed to him the cogency of Paley's celebrated clock analogy. "If the proportions and properties of land, sea, and air were not adjusted according to the reciprocal capacities of all to perform the functions required by each," he pondered, "why should we be told that He 'measured the waters in the hollow of his hand, and comprehended the dust in a measure, and weighed the mountains in scales, and the hills in a balance'?" (Maury 1874, 105).

Further instances of geographical teleology could be elaborated ad libitum. In this regard, Victorian geographical writing is hardly exceptional. Reinforcement for what might be called doxological science was readily forthcoming from such members of the British scientific fraternity as David Brewster, William Buckland, Michael Faraday, Hugh Miller, Roderick Murchison, Richard Owen, Adam Sedgwick, and William Whewell, not to mention the diverse authors of the *Bridgewater Treatises*, and from such American figures as Louis Agassiz, Alexander Dallas Bache, James Dwight Dana, Asa Gray, Joseph Henry, Benjamin Silliman, and Alexander Winchell. That their science was domiciled within a natural theology framework confirms the significance of teleology as a common context for their scientific endeavors.

During the past two centuries, natural theology as an enterprise has been subjected to attack from philosophers, scientists, and indeed, theologians themselves. As we have already noted, Hume and Kant mounted their epistemological critiques on the limits of reason and on the deficiencies of analogical argumentation; in turn Darwin's theory of natural selection subverted William Paley's schema by showing how organic adaptation to environment could have arisen by purely natural means; and in the twentieth century the Swiss theologian Karl Barth declared himself an avowed opponent of all natural-theology. Despite such criticisms, the natural-theology enterprise has continued to cast a long shadow over the heritage of geographical thought and practice—and on occasion with quite remarkable acuity.

Consider, for instance, the case of Gerben De Jong who, as recently as 1962, rooted his concept of geography as fundamentally "chorological differentiation" in teleological soil. Here, engaging in a Ritterian reappropriation, he spoke in Kantian sounding idiom of the region as "a form of thought" and went on to elaborate what he called a "chorological teleology." This was simply because "the chorological differentiation of the earth does not exist in and by itself, but is sustained by divine energies" (De Jong 1962, 62, 138). That this affirmation also carried moral conviction briefly surfaced in De Jong's further assertion that:

> although the dark shadow of the curse resting on the creation is clearly discernible in the antithesis in which the chorological differentiation of many regions has degenerated, the wealth implied in this earthly diversity has not been destroyed by the curse. On the contrary, owing to divine grace the harmony of the unity in the diversity in various regions has become manifest at every point of time (De Jong 1962, 138–39).

The philosophical underpinning of this *pronounciamento* had already been declared a few years earlier in De Jong's Kuyperian elucidation of "The Nature of Human Geography in the Light of the Ordinances of Creation." Here he grounded his exposition of geography as areal differentiation in Kuyper's concept of common grace and urged that the creation ordinance impelled the human race "to cultivate the richness of areal diversity of the physical earth and develop it into a richness of areal diversity of material culture." And here the normative impulse in De Jong's teleological imperative more clearly manifested itself: "the more this areal functional teleolo[g]y is present in a human geographic area, the more this human geographic area or region will answer its purpose which is founded in the ordinances of creation." The implications were immediate and he spelled them out in a quite remarkable way. Regions displayed both beauty and corruption, wealth and poverty in varying degrees; and to the extent that their holistic integrity was subverted either by jingoistic overdevelopment or by degeneracy "into similarity" they participated in "the consequences of sin." "Nationalist endeavors," he added, "is an example of the former, some forms of colonization and the communist world domination are instances of the latter" (De Jong 1957-58, 106, 109, 111).

De Jong's efforts to provide a Christian framework for the study of geography did not go unnoticed by others with similar intentions. Thus, Gordon Lewthwaite found his prescriptions, and indeed those of other Dutch Reformed thinkers—like the historian of science Reijer Hooykaas and the philosopher Herman Dooyeweerd, compelling. He too turned to a variety of biological and environmental arguments for teleology in pursuit of his conviction that, citing

De Jong, "the chorological diversity of the earth is one of the treasures of the creation." Accordingly, he likewise judged that "the healthy growth of areal differentiation can be stunted by the imposition of monotony or exaggerated into an unhealthy self-sufficiency, while the beneficial process of spatial interaction can also spread corruption and impoverishment." Diversity was thus "the norm of human geography" (Lewthwaite 1973, 181-82; see also Lewthwaite 1971).

That geography was, for much of its history, practiced within the framework of natural theology certainly does not mean that all its claims should be endorsed. To the contrary. Some of the practices associated with physico-theology are in need of repudiation, for an overweening belief in the providential order could be used, in theodician fashion, to justify the worst forms of racial injustice and social inequality. The now unsung verse of Cecil Frances Alexander's hymn "All Things Bright and Beautiful" is illustrative: "The rich man in his castle, the poor man at his gate; God made them high and lowly, and ordered their estate." Here the religious justification of the established order is only too easily unveiled. Moreover, as historians such as the Marxist R. M. Young have shown us, such religious theodicies were easily translated into scientific idiom in the form of naturalized theodicies. Young's thoroughly Marxist conviction is thus simply that science and religion are both socially sanctioned ideologies. In his reading, the nineteenth-century debate about "man's place in nature" is fundamentally the story of the substitution of a religious theodicy by a scientific one; in both, the status quo is legitimated, first by talk of divine law or natural theology, and latterly by the language of natural law or natural selection. Science, like religion, merely acts to support principles of social conformity. Indeed when one looks at the rhetoric on the lips of leading scientific publicists like Huxley with talk of "lay sermons," "scientific priesthood," "the church scientific," and "molecular teleology," Young's arguments seem to have much to commend them. Science here seems to occupy the role of a naturalized natural theology. Values are no longer drawn from the supernatural realm but are earthed in the all-too-mundane world of nature. And this naturalization of values is achieved through the sacralization of science (Young 1985).

Nor has geography been immune to such transformations. Geographical natural theologians, like Guyot, had for long enough been justifying racial inequalities by reference to the will of the Creator. Such sentiments were easily translated into secular scientific discourse (see Livingstone 1991). In her recent reevaluation of what she calls Griffith Taylor's "racial ecology," Nancy Christie contends that, in contrast to other evolutionary theorists, Taylor "eviscerated all moral concern from his scientific endeavors." As she tellingly puts it: "Natural law operated in Taylor's world view with Paleyite regularity, but Nature was technocratic, elitist, and efficient, and its laws were to be administered by a new priesthood of expert geographers and social scientists, who could interpret the predictive patterns of nature for social engineering and national planning" (Christie 1994, 432, 439).

Having dwelled on some of ways in which teleology could subserve the interests of repression and injustice, it would be mistaken to reduce the natural-theology tradition to nothing but an expression of political manipulation and ideological self-service. Its historical value in addressing issues of organic adaptation to ecological niche, for example, reveals its empirical fertility. At the same time, the enthusiasm with which contemporary scientists and philosophers—such as John Polkinghorne (1986), Arthur Peacocke (1985), Paul Davies (1984), and Richard Swinburne (1979)—have embraced various renditions of the design argument displays something of its continuing intellectual vitality. Given these more recent moves, moreover, it seems appropriate to turn now to some philosophical matters intimately bound up with this whole project.

Natural Theology and Foundationalist Epistemology

I believe I have said enough to indicate just how influential the "teleological imperative" has been on the evolution of the geographical enterprise. But embedded within this tradition are crucial epistemological issues that have been at the center of recent debate over questions of knowledge, belief, and rationality. In a nutshell, I think there is much to be said for the view that the natural-theology enterprise was profoundly implicated in the establishment of modern epistemological foundationalism. In order to get a handle on these matters, however, I believe it is crucially important to distinguish between two varieties of natural theology that, in all likelihood, have been elided in the cases I have thus far laid before you.

These two versions of the design argument can be summarized as the argument *from* design and the argument *to* design (see McPherson 1972). The former—the argument from design—constitutes an argument postulating that instances of design in the world provide grounds for belief in the existence of God. In other words, evidences of design are to be used as an argument for a divine architect. The argument from design is thus one of the classic "proofs" for the existence of God. By contrast the argument to design presupposes that the world is the creation of a supreme divine agent and that it will therefore provide evidence of his design. In this case, God's existence is not the conclusion of an argument but its presupposition. The former is a philosophical claim whereas the latter is a confessional proclamation.

In his account of "The Reformed Objection to Natural Theology" Alvin Plantinga (1982, 187) notes that Reformed thinkers, in general, have "taken a dim view of this enterprise." And this for the reason that Calvinists do not think that belief in the existence of God should be based on proofs or arguments. In Plantinga's reading, natural theology is inescapably embroiled in epistemological foundationalism, that is, the assertion that in any rational noetic structure (to use his terminology) knowledge must ultimately be grounded in propositions that are either self-evident or evident to the senses. In similar fashion, Wolterstorff (1984, 28) defines the foundationalism that has dominated the Western tradition in the following terms: "A person is warranted in accepting a theory at a certain time if and only if he is then warranted in believing that theory belongs to genuine science." Classical foundationalism, of course, has come under increasing (some would say defeating) pressure in recent times. Not only does it seem to be a self-referentially incoherent thesis, but its exclusion of all sorts of beliefs that are clearly held quite rationally—such as memory beliefs—has meant that many have come to reject it as an adequate account of warranted assertibility (see Plantinga 1983). Whether a weaker form of foundationalism—one incorporating a wider range of good grounds for rational belief—is to be adopted, or whether the whole foundationalist enterprise had better be consigned to the trash can of philosophy is an open question. Plantinga seems to want to retain a modified foundationalism while Wolterstorff is more prepared to jettison it altogether. Whichever, in Plantinga's reading, the Calvinist objection to natural theology stems from the conviction of that tradition that belief in God's existence is not to be inferred from other more foundational propositions. To the contrary, God's existence is itself a foundational claim, or as he puts it, properly basic: "Among the central contentions of these Reformed thinkers, therefore, are the claims that belief in God is properly basic, and the view that one who takes belief in God as basic can *know* that God exists" (Plantinga 1982, 195).

If this reading is in the neighborhood of a correct analysis, two implications for geography are to be recorded. First, it might seem that the whole tradition of geographical natural theology has been a philosophically and theologically mistaken enterprise. But this is to do insufficient justice to the distinction that I elaborated at the start of this section. In fact it seems to me that very few who embarked on the task of illustrating design in nature were operating in a foundationalist mode at all; many, perhaps most, were actually pursuing a confessional teleology of the sort that since God exists, there will be traces of design in His world.

Secondly, if indeed it is the case that classical foundationalism is antipathetic to the spirit of the Reformed tradition, then the collapse of the Enlightenment project should not strike the terror to the hearts of contemporary Christians that many have thought it should. "As the Reformed thinker sees things," Plantinga asserts (1982, 198), "being self-evident, or incorrigible, or evident to the senses is not a necessary condition of proper basicality." This suggests that there are resources within the heritage of Western Christianity for coming to terms with postmodern thought forms. Allow me to illustrate this by one example. In the wake of the fragmentation of positivistic foundationalism and the unmasking of the rhetoric of scientific neutrality, scholars have turned to the central issue of trust in the making of knowledge. Steven Shapin (1994), for example, argues that all of us, scientists and laity alike, hold the bulk of our knowledge "by courtesy." Knowledge requires trust, and trusting, he reminds us, is a form of faith indispensable to "the holding and growth of scientific knowledge." Most of what we know about the natural world we believe on somebody's authority. The foundationalist rhetoric of science's positivistic champions is thus revealed to be mere epistemological bravado. But this is hardly a new insight, is it? Already in the late fourth century, St. Augustine (1961, 117) had written: "I began to realize that I believed countless things which I had never seen or which had taken place when I was not there to see—so many events in the history of the world, so many facts about places and towns which I had never seen, and so much that I believed on the word of friends or doctors or various other people. Unless we took these things on trust, we should accomplish absolutely nothing in this life." Socializing epistemology, it seems, has a long history, and had we been attentive to our own heritage, the shocks of positivist and then postpositivist assaults might have been more easily absorbed.

Historiographical Retrieval

The case study that I have sought to lay out does, I think, go some way to exposing the weaknesses of a Whiggish historiography that persistently reads the geographical tradition as a progressive story toward modern secular orthodoxy. Not only has this impaired our history, but it also cut off from access historical sources for grappling with contemporary problems. I believe that I have illustrated the vitality of natural theology in geography's history and have gone a little way toward recovery. But this does not exhaust the range of topics that are in need of reconsideration. We might refer, for example, to a growing acknowledgment of the mystical and sublime on certain aspects of the geographical tradition (see Cosgrove 1990; Matless 1991). Allow me here, however, to hint at just three further ways in which we might engage in some historical reappropriation.

The positivist historiography under which the writing of geography's history has suffered, I believe, has undervalued the role of religious conviction in the evolution of the discipline. Consider, for example, the significance of the missionary enterprise in the growth of geographical knowledge. Its role has generally been neglected or, when it has surfaced, been subjected to the facile, monistic critiques of a faddish postcolonialism. The beginnings of interest in the subject just now, moreover, have much to do with what Lamin Sanneh (1987) has dubbed the "Western Guilt Complex." While there are doubtless many grounds for the attribution of missionary guilt, the work of Adrian Hastings (1996) and Sanneh himself (1993) on the connections between mother-tongue cultures, Christian mission, and senses of identity has begun to redress something of a balance. Unfortunately I am not in a position to present material on the missionary production of geographical knowledge; but I can certainly work by analogy with other sciences. In a recent collection of essays on science in the Pacific, for example, several contributors turned attention to the role of missionaries in the production of psychological, anthropological, and ecological knowledge. Gunson, for instance, concluded his assessment of "British Missionaries and Science" with the observation:

Although many articles on natural history, geography, and ethnography were published by missionaries in religious periodicals for the edification of the public at large, other, more detailed articles were contributed to the journals of learned societies and so made a contribution to scientific advance. From the late 1860s there was a steady flow of specialized articles ranging from vulcanology to linguistics. . . . Several of the missionaries played key roles in extending contemporary knowledge of the natural world (Gunson 1994, 306).

Again, John Stenhouse, in a remarkable reassessment of Darwinism and New Zealand politics in the same period, shows how Darwinian "enlightenment" was commandeered to underwrite the settler's imperialist attitudes toward the Maoris, in contrast to the humanitarianism of missionaries with evangelical convictions. In the New Zealand context, it was a case of "Settlers versus Maori and Missionaries" (Stenhouse 1994, 398). Of course this is not to prejudge the issue as to how an examination of the role of missionary enterprises in the history of geography would turn out; it is to say that the topic is worthy of consideration.

Similar conditions pertain, I think, in discussions of the idea of landscape. In a recent article, Robert Mayhew has made a good case, in opposition to the prevailing Marxist orthodoxy, for the view that the idea of landscape in the eighteenth century was firmly domiciled in what J. C. D. Clark (1985; 1989) calls "denominational discourse." If Mayhew is to be believed and landscape thinking in this period was embedded not simply in a Christian framework but in a variety of denominational matrices, this would certainly open up new lines of inquiry while at the same time further contributing to the undermining of "the traditional historiography of a secularizing Enlightenment" (Mayhew 1996, 454-55). Such work contributes at once to the task of historical revisionism and—perhaps even more important—of uncoupling the presumed universal determination of discourse by the socioeconomic. As Mayhew (1996, 467) concludes: "In its sensitivity to the languages of debate in historical and geographical context, and its awareness that these display a dynamic of their own which cannot be simply mapped onto social-historical realities, revisionism makes distinctions of great importance to students of environmental history and discourses about that history."

A third arena in which greater historiographical subtlety is required, I believe, is in the interpretation of ideological import of environmental determinism. Richard Peet's (1985) diagnosis, which reduces the entire environmentalist project to issues of racist and capitalist manipulation, is simply much too coarse grained. In fact there were many occasions when climatic determinism was resorted to in order to support racial equality and confirm the unity of the human constitution. I have recently been engaged, for example, in an exploration of the role of such themes in the early American Republic. Here I have discovered that Samuel Stanhope Smith, President of Princeton, Presbyterian clergyman, advocate of Enlightenment, and the "father of American anthropology," used climatic theories of racial modification to underwrite a common human nature, to minimize ethnic differences, and to support a form of cultural relativism. Once again stereotyped historiography does not serve us well.

These hints, such as they are, at historiographical retrieval go no more than a little—but at least a little—way toward confirming suspicions similar to those articulated for sociology by John Milbank. For him "the most important governing assumptions" of modern secular social theory "are bound up with the modification or the rejection of orthodox Christian positions" (Milbank 1990, 1). Absorbing certain diagnostic features of postmodern antifoundationalism, Milbank goes on to mobilize these in his revisiting of a sequence of key moments in the emergence of modern secular rationality; what he finds—again and again—is the contingent and the arbitrary masquerading as the objectively obvious closure of epistemic negotiations. These inquiries, in turn, serve to reinforce Milbank's conviction that social theory has persistently acted "as a secular policing . . . to ensure that religion is kept, conceptually, at the margins" (Milbank 1990, 109). Where it does surface, religion features only as a dependent variable explicable in terms of, and reducible to, societal functioning, an ideological will to power, or some such. The narrative I have woven above likewise suggests that modern geography has routinely colluded in a similar historical and theoretical maginalization of the metaphysical.

Alternative Agendas

If there are good historical grounds for identifying key metaphysical issues at the heart of the Western geographical enterprise, and if there are solid philosophical reasons for the legitimacy of having religious convictions as properly basic control beliefs, then there are significant implications here for the practice of a Christian geography. For one thing, it implies that Christian geographers, Jewish geographers, Islamic geographers, not to mention Marxist geographers, feminist geographers, positivist geographers, and so on, are entirely within their epistemic rights to pursue projects in keeping with their core beliefs. In turn, this further implies that a pluralism in the academy is both inevitable and desirable. At the same time, it reaffirms the need to reject the cultural imperialism that, operating under the guise of academic neutrality, polices the discourse with antimetaphysical prejudice. For too long we have allowed the established norms of the academy to dictate the cognitive content and conceptual frameworks of the enterprise.

Let me work here by analogy with philosophy, drawing on the contributions of Alvin Plantinga. For a good deal of this century, philosophical theology was held captive to the positivist principle of verifiability according to which propositions only made sense if they could be tested by empirical observation. This meant that assertions about God were held to be literally nonsense, meaningless, because they failed to live up to an empiricist criterion of meaning. The debate revolved around whether it even made sense to claim that there is such a person as God. As things turned out, the verificationists really had no good arguments to sustain their own position. For a start, their own verification principle was not itself an empirically testable claim. Moreover the principle self-evidently ruled out all sorts of bona fide claims whose meaning was never really in doubt. Plantinga concludes from this episode that "Christian philosophers should have displayed more integrity, more independence, less readiness to trim their sails to the prevailing philosophical winds of doctrine" (Plantinga 1984, 258). I suggest the same is true of modern geography. We have been too ready, for example, to stand by and watch the dissolution of the human agent, the self, and the author by those of deconstructionist stripe. Instead we might have culled the riches of our own tradition to reaffirm the dignity of the human subject—a subject mirroring the *imago dei*, and not Marx's flotsam and jetsam of the mode of production, Darwin's trousered ape, Freud's bundle of sex drives, Nietzsche's (and therefore Foucault's) subjectivist negation of the author and the substantial self, and so on. When we are grappling with just what it is to be human person, as Plantinga (1984, 265) puts it, the "first point to note is that on the Christian scheme of things, *God* is the premier person, the first and chief exemplar of personhood. . . . How we think about God, then, will have an immediate and direct bearing on how we think about humankind."

The same situation, I venture to suggest, pertains at a time when issues of representation, construction, and imagining are all the rage in human geography. For many, the underlying conviction is simply that everything is just representation, and that no judgments can be made about adequacy or inadequacy, accuracy or inaccuracy, truth or falsity. In this environment, surely we will want to say that certain representations of people and places, or races and regions, ought to be repudiated precisely because they are *un*authentic depictions or characterizations of the human subject and its habitus. "When we stand within the moral outlook of universal and equal respect," Taylor (1989, 67) argues, "we don't consider its condemnation of slavery, widow-burning, human sacrifice, or female circumcision only as expressions of our way of being, inviting a reciprocal and equally valid condemnation of our free labour, widow-remarriage, bloodless sacrifice, and sex equality from the societies where these strange practices flourish." Certainly some do seem to think in these terms on account of postcolonialist sympathies; but this is surely strange since the postcolonial impulse itself is actually a consequence of taking universal respect seriously! Of course the idea of universal justice—that all humans should be treated equally irrespective of race, class, gender, culture, religion and so on—has had to be fought for. Arguments have had to be marshalled to convince interlocutors of the oppressive character of certain characterizations and the worthiness of their alternatives.

Now, to be sure, there are dangers in this way of thinking—Foucault's explorations or, perhaps better, unmaskings of the insidiousness of modern surveillance over against its antecedents is itself enough to establish this. What is seen as fulfilment or even liberation can turn out to be yet another more cunning form of enslavement. Yet, I believe we should be suspicious of the movement of European high culture toward the neo-Nietzschean nihilism that has done so much to negate "all horizons of significance." For myself, I think it is only when we retain notions of authenticity, dignity, and justice, when we hold on to ideals of "better" and "worse," "true" and "false" that we can say, for example, that Victorian opposition to universal suffrage was based on an inadequate concept of the human nature of half the world's population. Or again, adopting hiring practices that ensure that women get their fair share of jobs constitutes a richer picture of the human condition than those undergirding polygamy, or purdah, or suttee, or killing "fallen" women for the sake of family honor.

It seems to me that in both these cases—and doubtless many others—the Christian thinker is clearly within his or her rights to approach them with their own presuppositions in hand and not to assimilate to the currently fashionable or politically correct tropes of the secular academy.

Allow me to end these reflections where I began, with the words of W. M. Davis:

> It has often been claimed that the marvelous progress of civilization in Europe has resulted from the adoption of the Christian religion; and it would be indeed gratifying to know that the refined ethics of Christianity have really brought about that progress. But in view of the dependence of European civilization largely on the advance of science, which did not begin until the revival of learning centuries after the adoption of Christianity, and in further view of the persistent opposition with which the organized forces of Christianity so generally resisted the advance of science . . . the claim that European civilization is a result of Christianity can be allowed only in part, perhaps only in small part (Davis, quoted in Chorley, et al. 1973, 773).

Here, I suggest, to expound is to expose.*

* I am most grateful to David Ley for a number of helpful observations on an earlier draft of this chapter.

Chapter 2

Postmodern Epistemologies: Are We Stuck with Our Relatives?[1]

David Ley

WE ARE TOLD that a major restructuring of epistemology, our means of knowing, is underway in the contemporary university. In the humanities, this is old news, as what in the broadest sense we might call postmodern epistemologies are established and are on the way to becoming a new orthodoxy. In the social sciences, the initial skirmishes are long over, and younger faculty and graduate students in particular are pressing to secure further advances. Even in the physical sciences there is disquiet, an awareness that truth is perhaps not what it used to be, at least in the judgment of many critics. Both theological and social-scientific response to these developments should be, I think, properly ambivalent. While some defenders of the orthodoxy of modernism see in postmodern epistemology only the anarchy of postmodern relativism, to which they couple the antinomianism of postmodern culture, others see a toppling of a totalizing worldview that claimed entirely too much for itself and, in the habit of all totalitarianism, had excluded all other wisdom.

These broader currents achieve a sharp focus in our own province of geography, which in its catholicity represents an unusual convergence of the humanities, social sciences, and physical sciences. The revolution that David Lowenthal (1961) inaugurated over thirty-five years ago in his seminal reflections on geographical epistemology continues to unfold and destabilize any simplistic view of knowledge. While positivism, structuralism, and to a considerable degree, realism, have insisted in the intervening years upon sustaining the idea of a fully knowable geographical world and a privileged vantage point for the researcher, the spreading influence of perception studies; humanism; and more recently deconstruction, feminism, postmodernism, poststructuralism, postcolonialism, and the sociology of scientific knowledge— have all exercised growing subversion on such a comfortable universe by arguing that reality is a plural not a singular term, that the researcher is fully implicated in the production of knowledge, and that the dissemination of knowledge is also the deployment of authority.[2] There are of course significant differences among the representatives of this literature—in particular, Lowenthal's emphasis on the individual and psychological models has given way to a concern now with society, history, and the social construction of reality— but all of them would concur with the theologians Richard Middleton and Brian Walsh (1995) that *Truth Is Stranger Than It Used to Be.* So this is the terrain I wish to cover in preliminary form in this chapter. Note that the focus is not on postmodern movements in the arts or architecture, nor the character of what some call postmodern urbanism, or a postmodern society. These are important matters about which much has been written, but they are matters for another occasion (cf. Ley 1996). What I hope to do here is to raise some issues for discussion, to start a conversation, in the realm of epistemology. In practical terms, epistemology cannot be separated from worldview, and the concept of worldview immediately alerts us to the theological precedent for such discussion, particularly in the Reformed Protestant tradition, which we shall meet shortly. The conversation, then, will bring together three voices: advocates of postmodernism, human geographers, and Christian theologians. There are three major themes in my commentary. First, I wish to review the critique of Enlightenment epistemology offered by postmodern thought. Second, I shall suggest some additional contributions that this position offers. Third, I shall confront the hardest issue. In engaging postmodernism, how do we get around the difficult problem of relativism? An epistemology that argues that reality looks different according to the position of the observer has, inevitably, a problem in resolving the different views of different observers. Are we simply stuck with our relatives? And how then do we deal with the claims of truthfulness of the Christian gospel, and in particular the declaration of Jesus to be the way, the truth, and the life? How is Christian truth to be accommodated with postmodern skepticism? My engagement with these questions will take the form of both argument and confession; it is implicit in what follows that such a posture is a fuller, more reflexive, and indeed more honest mode of presentation. Whether in science or religion, we bring to any project both reason and faith. As recent work in the sociology of knowledge, including geographical knowledge (Livingstone 1992), has eloquently shown us, the disembodied intellectual tradition exemplified by Descartes was always contained within an often unacknowledged nexus of material, social, and, yes, spiritual relations.

Challenges to Enlightenment Knowledge

I will begin by referring to Nicholas Wolterstorff's (1995) recent essay "Does truth still matter? Reflections on the crisis of the postmodern university." Wolterstorff, professor of philosophical theology at Yale, writes from within the Reformed tradition. His essay examines the nature of postmodern challenges to scientific modes of knowing—challenges, he argues, that are directed not just at the Enlightenment period but at the whole history of Western learning, what he calls the Grand Project dating from Plato and the classical tradition of Greece and Rome. Against this tradition are arrayed the ranks of perspectival learning, such as current viewpoints on feminism, deconstruction, postcolonialism, social constructionism, neopragmatism and so on—viewpoints, to be sure, that have their own emphases but have in common the fact that they self-consciously represent a view from somewhere—perspectival or positional knowledge that rejects what Richard Rorty (1991) has called a God's-eye view and what Donna Haraway (1996) dismisses as the god-trick of unmediated knowing. They deny the claim that knowledge is a full mirror and correspondence of a reality that is fully accessible to anyone. For the purposes of this discussion I will follow Wolterstorff and group these perspectival viewpoints as representative of postmodern epistemology, while acknowledging that they are far from homogeneous and contain significant variation among their parts.

For present purposes what they have in common is more relevant. They all offer substantial criticism of the Grand Project. That project in its epistemological guise, notes Wolterstorff, claimed to be a voice from nowhere, free from the distractions and biases of everyday life. The intellectual mind removed these opinions, together with its overcoat, as it entered the library or the laboratory. It pursued thinking that was objective and value-free, engaging an accessible and necessary reality and pursuing a project that was visionary and cumulative. A dominant image is departure, leaving behind the everyday to enter a world that was dispassionate and rational, and moreover, promised to be a better place, closer to immutable truth than the flotsam and jetsam of everyday life. In this rarified space "one lets one's reason put one in touch with necessary truths" (Wolterstorff 1995, 22).

We recognize this view as indeed the master narrative of academic epistemology. It was the view that was drilled into me as a graduate student as the scientific method. Notice that definite article and the totalizing claims it makes for itself. But it was a view I found myself starkly at odds with. Its assumptions to get at the really real, the necessary—such assumptions in human geography as fully rational economic behavior (remember the unproblematic world of *rational economic man*, each term now commonly greeted with quotation marks or *sic* in parentheses) and a geographical landscape with all of the diversity of a billiard table—seemed facile. The scientific method's avoidance of real places and preference for abstract spaces seemed ingenuous. Worst of all, it limited the kinds of problems we could encounter. The method became the message. If it could not be observed and counted, it could not be studied. And so we addressed problems that were often jejune. And here was the deep irony, for as we flexed our muscular numeracy, and keyboarded packs of computer cards, outside, on our very campus in 1968, fires were lit and the National Guard patrolled. The issues of Vietnam, environmental catastrophe, civil rights, and urban rioting; our methods had nothing to say to any of them. These issues were too particular, too complex to be domesticated and simplified. They could not be studied, only talked about with passion once we left the laboratory. In the laboratory, we were busy figuring out the statistical probability of the numbers of migrants from Alabama who would end up in each of the largest cities of the United States. The results of such disembodied empiricism were sometimes downright hilarious. As government quickly turned its funding lens to disaffected youth in the late 1960s, social scientists in Ontario devised an alienation index, which to everyone's relief showed the disaffected occupying a tail end of a normal distribution. With the problem of radical criticism appropriately quarantined, the next task was to establish a social control index, though in a parody of contextless empiricism, researchers admitted: "Just what it measures is difficult to determine. . . ." (Byles 1970).

This is a personal diversion to be sure, but a diversion that explains my own restlessness with *the* scientific method. It is not the only objection of course, and Wolterstorff expresses a second: The view from nowhere is always a view from somewhere. We cannot divest ourselves of our culture, our socialization, or our personality when we enter the university. "What transpires in the academy," writes Wolterstorff, "is not an alternative to everyday life but an intensification and extrapolation of it" (1995, 26). In the history of the academy, the view from somewhere, as we are frequently reminded, is the view from a middle-class, Euro-American, and male perspective. It is the perspective of a special interest that claims universality for itself. Its knowledge is therefore a representation of the real, (doubly) partial knowledge, and not the fulsome knowledge of the thing itself that we are taught to expect. Within human geography Deutsche's feminist critique of Harvey's *Condition of Postmodernity* is most persuasive (Harvey 1989; Deutsche 1991). The sweeping and heroic range of Harvey's perspective is brought to ground as an unacknowledged patriarchal project, not only partial but also partisan.

Consider also the debate around competing representations of 1492, a pinnacle of achievement and the opening of new frontiers according to one perspective, the experience of slavery and genocide for another. Or consider the very term, the voyages of discovery, to illuminate the partiality of that Eurocentric perspective. Remember too that John Locke, a founding father of the Enlightenment version of the Grand Project, despite his philosophical rationality, had the most disparaging things to say about North American aboriginal peoples —opinions that today would very quickly land him in court. Or if our bent is toward more rational models of theology, consider the infelicitous views of John Knox toward women. In each instance, an apparent view from nowhere quickly reveals the socialized interests of those who have shared in its construction.

There is a third and related issue of importance to raise here. In his seminal, if difficult, book, *Theology and Social Theory,* John Milbank (1990) has observed how the supremacy of rational epistemology marginalized other ways of knowing. Indeed it banished them, as the scientific method of observation and deduction claimed the whole field of intellectual knowledge. Milbank uses postmodern theorists to shrink down those claims, to represent scientific epistemology as an exercise in regional, not universal, knowledge. Its claims to universalism, notes Milbank, are metaphysical claims, and reveal that it has subverted religion in order to become a religion. In other words, social science's critique of metaphysics is a new metaphysics laying claim to a universal empire (Milbank 1990, 105). Religious experience is thereby policed, contained within the inadequate categories of a particular form of empirical knowledge. But such policing requires a superior epistemological vantage point, a God's-eye view, and claiming such a vantage-point is metaphysical. Therefore, Milbank concludes, "every secular positivism is revealed also to be a positivist theology" (1990, 139). This argument has been exemplified by Neil Smelser in the particular case of economics, the most prestigious of the social sciences: "The constructed economic paradigm contained all the major elements of a religious system: a worldview or cosmology; a system of legitimizing values, orderliness and coherence as a belief system; more or less derivable moral implications; and factual claims about the empirical world" (Smelser 1995, 81). Trevor Barnes (1996), in a bold and innovative deconstruction of the presuppositions of economic geography, has added a devastating rider. Although proponents of the scientific model in economic geography made the grandest claims for their method, they failed to follow its rational protocols in their own work. Rather than universal, writes Barnes, theirs too, was a form of local knowledge, mired in personality, convention, and the spirit of the times. I am reminded of the wise observation of the phenomenologist Alfred Schutz, that although everyday life, the taken-for-granted world, is routinely banished from the scientific establishment, it always secretly returns to exact its revenge.

In a postmodern academy, then, the assertion of epistemological high ground, the God's-eye view, is presently inadmissible. An important consequence of the destabilization of some of the certainties of modernity is that it makes space for other ways of knowing, including the religious. One recent collection, edited by a lecturer in English at Cambridge and a professor of cultural studies at Trent University, is entitled *Shadow of Spirit: Postmodernism and Religion* (Berry and Wernick 1992). The introduction cautiously enters long-forbidden ground, observing that it is only "in the last few years that we have begun to elaborate 'other' ways of thinking that alterity which philosophy consigned to the marginality of darkness. Probably one of the most unexpected results of this changed perspective has been to revive interest in those once-tabooed aspects of 'otherness' which can broadly be termed spiritual or religious" (Berry 1992, 3). Clearly the postmodern academy offers opportunities for the recuperation of religious frames of knowing from the scrap heap of modernity.

Further Applications of Postmodern Epistemologies: Church History and the Biblical Record

We are now edging our way toward the specter of relativism, the equivalence of all possible worldviews, but I do not want to draw back yet. I want to suggest that this view of knowledge as socially constructed does not only prove enlightening as a critique of the Grand Project, or in presenting feminist or postcolonial insights, or in establishing the unacknowledged conventions of economic geographers but may also illuminate aspects of the history of the church and is even sustained by events from the biblical record itself. How often in ecclesiastical history have sectarian groups mistaken their own social constructions for an absolute revelation of truth. These tight bonds of dissenters produce almost a perfect example of Richard Rorty's discursive subcultures;

with repeated interaction and a common consensus they all too readily convince themselves of their own omniscience and the error of others. Quite literally they claim a God's-eye view, and, from this vantage point of absolute truth, they pour invective on others. A postmodern epistemology would point to the social construction of their knowledge, their claim to omniscience, and the error of presumption that accompanies it.

Paul's letters to the Corinthians have a good deal to teach us in this area, and to my mind, they suggest that Rorty is to be taken seriously. The Corinthians, remember, were part of that classical Greek tradition that Wolterstorff identifies as the origin of the Grand Project of the Western university. In the first chapter of his first letter, Paul begins by acknowledging their knowledge and their search for wisdom, their respect for the wise man, the scholar, "the philosopher of this age" (1 Cor 1:20). But like many sophists since, their quest for knowledge was shot through with prejudice. In place of respect, quarreling and division marked their mutual intolerance the one for the other as they debated their cases. To this factionalized, proud, and defeated congregation, Paul goes on to show that "the most excellent way" (1 Cor. 12:30), better than human knowledge, better even than inspired spiritual knowledge, is love. This advice could profitably have been accepted by Luther and Zwingli in their bitter exchanges, or in many similar cases in church history.

A postmodern epistemology elicits a certain modesty that there is much to be known about any topic and that we cannot know all of it. If our knowledge is partial, then we should be cautious about the claims we make. To make stronger claims is to pretend an omniscience that is intellectually untenable and spiritually presumptuous. In the first letter to the Corinthians, Paul moves from the human knowledge extolled by the Greeks, to spiritual knowledge, to love. This is also an ethical progression, from the good, to the better, to the best. As he moves to the best, the way of love in chapter 13, he contrasts it with inspired knowledge. What does he say about that knowledge? "We know in part and we prophesy in part. . . . Now we see but a poor reflection [through a glass darkly]; then we shall see face to face. Now I know in part; then I shall know fully, even as I am fully known" (vv. 9-12). Partial knowledge is the best we can hope for, even for one with the deep spiritual insight of Paul. Whenever we acknowledge partial, indeed positional, knowledge we have substantial biblical precedent.

There has been widespread recognition, both empirically and philosophically, of the bases for, and the implications of, our inevitably incomplete accounting of knowledge. Following Lowenthal's (1961) seminal insight, the perception research of the 1960s and 1970s, including the work on mental maps and cognitive mapping, underscored the partiality of our geographical knowledge, and that insight has been consolidated on an epistemological plane by a number of contemporary authors. "Feminism" writes Haraway (1996, 121), "loves another science: the sciences and politics of interpretation, translation, stuttering, and the partly understood." Here "in the politics and epistemology of partial perspectives" (1996, 117) is the ground for a critical feminist agenda. Addressing the problem of representation in anthropology and other cultural fields, Clifford (1986, 7) advocates "partial truths," a "rigorous sense of partiality" that similarly rejects an unproblematic realism that disregards the social and cultural contexts of the production of knowledge.

There is a second area where postmodern epistemology may be illustrated by incidents from the historical record of Scripture. Rorty takes the consequences of perspectival knowledge to its strong conclusion. Social life consists of linguistic communities who are defined by their ethnocentrism. Truth becomes a group-specific construction; it is their take, their perspective on the world. Truth, writes Rorty, is "entirely a matter of solidarity" (1991, 32). This view seems an extreme one and has been much challenged, but does it not throw some illuminating light on the relations between Jesus and the religious elite of his day? Surely they belonged to different discursive communities. For the Pharisees, truth was a way of seeing resulting from a socially constructed reality, a matter of solidarity; Jesus was interpreted within a set of intersubjective conventions. Those conventions did not allow the Messiah to come out of Galilee, they did not permit him to take on the form of servanthood, and they did not enable him to transcend the religious law. Any tampering with their conventions was a threat to the solidarity of their community and to the deployment of their authority. And so they became as Jesus observed "blind guides" (Matthew 23:16). They were fully contained within the community; they could not communicate across its boundaries. Jesus was speaking a language they could not understand. John's gospel in particular highlights the failure of the Pharisees (like later skeptics) to comprehend his pronouncements or his parables, a failure that was frequently poignant in its irony.

Not only does a perspectival epistemology have a valuable criticism to make of the Grand Project of the Western university, it is also illuminating in understanding social life, including church history. Indeed, we can see points of application in the biblical record itself. But does this recognition of partial knowledge come at too heavy a price? Clifford (1986, 6), seemingly in the same way as Rorty, takes the implications of constructionism a step further. Ethnographers are "tricksters," their interpretations are "fictions," but this is not "the banal claim that all truths are constructed." Rather, Clifford (1986, 7) seems to harbor a strong perspectivalism "more Nietzschean than realist or hermeneutic [where] all constructed truths are made possible by powerful 'lies' of exclusion and rhetoric." Is such partial knowledge therefore only self-referential, no more than a language game?

Yes . . . But What Then Is Truth?

At this point, as I move from the second to the third step in my argument, it may well seem that I have dug myself into a deep hole. For while there are important insights as I have tried to show in perspectival epistemology, it is time now to raise the objections, the yes . . . but responses that limit overhasty endorsement. A major objection for present purposes is the view of truth that is limited to its status as a product of a discursive community and with no greater validity than that. Beliefs for Rorty are derived from habits, repeated actions that emerge from adaptive relations with a community's environment and are true if they are coherent in terms of the community's adaptation and purposes. Clifford (1986) in advocating a postmodern anthropology makes more explicit what may be at stake in terms of the Nietzschean will to power in such socialized knowledge. No less troubling are Maffesoli's views on what he calls the neotribalism of contemporary societies where the ethical dimension is relative to the tribe's customs, "a vessel for the collectivity's emotions and feelings," while justice by the same group-centered principle "is subordinate to the experience of closeness; . . . abstract and eternal justice is relativized by the feeling (whether hate or love) experienced in a given territory" (Maffesoli 1996, 20, 17). The Christian views of truth, ethics, and justice are much stronger not least because these terms are so central to the church's self-description. The Christian view of truth scandalously claims to be more than local knowledge. When Jesus states, "I am the way, the truth and the life," he makes it clear that this is a message not for a small community, not for one or, for that matter, twelve tribes, but a message with relevance across space and time.

But now we encounter two conundra. First, in acknowledging the power of postmodern epistemology, how does one get away from the argument that Christianity is just one more perspective in the large menu of possibilities arrayed before us, and as such should be practiced quietly inside the bounds of its discursive community? And if one surmounts this argument, then the second conundrum appears. How does one then address the succeeding challenge that if Christianity claims to be more than another perspective, then it is imperial and insensitive, a colonizing project that knows no limits and respects no barriers; in short another Euro-American project where claims to knowledge become vehicles to power and oppression?

Rorty's pluralism has been challenged normatively because it does not provide consistent ethical grounds for refusing one perspective against another. To this I would add an unobserved historical objection. His neopragmatism draws its inspiration from the pragmatism developed by William James, John Dewey, and others in the United States earlier this century. That work was itself an important influence on the significant school of urban sociology, well-known to geographers, that was established in Chicago in the decades up to 1939. As these scholars looked at the multicultural face of Chicago, they identified a plurality of social worlds, internally coherent, but effectively noncommunicating with each other, the same fragmentation that Rorty too has identified. But the verdict of the Chicago School was that a society of social fragments, contained and separated, was in an acute condition of social disorganization, and in the 1930s, in particular, they spent considerable time trying to devise how a new model of organization, a new master discourse, could emerge in the city to tie the disparate discursive communities into a larger unity. In other words, the first time around, the pluralism of distinctive cultural communities identified by the pragmatists was thought to be dysfunctional. Does this assessment continue to be persuasive?

Given this historical object lesson, how might one respond theologically to the challenges of postmodern epistemology? Milbank (1990) identifies three possible responses. First, one may dismiss postmodernism out of hand and return to the Enlightenment postulate of an uncomplicated realism. This is a position that Milbank rejects because it simply ignores criticism rather than addressing it; moreover it returns the study of religion to the iron cage of secular social science. A second option is to take on board the postmodern program enthusiastically, but typically this leaves an impoverished spirituality that, in its anxiety not to offend, retains little more than a generalized mystical moment. A number of the essays in Berry and Wernick's *Shadow of Spirit: Postmodernism and Religion* (1992) fall into this category. The third option is to try to work creatively with postmodernism, and this is Milbank's strategy as he seeks to outflank the nihilistic tendencies of deconstruction in particular. Before leaving Milbank, I want to acknowledge also his response to the second conundrum, that of the potentially imperial nature of religious claims that refuse to be limited to the status of another perspective. His argument is that Christianity is not another case of power disguised as knowledge. It therefore negates the argument of deconstruction that finds power and oppression at the end of every rainbow. Instead Milbank sets out the Christian story as one concerned with the pursuit of servanthood, virtue, and peace. To this Pasewark (1993) adds the observation that a Christian view of power is one that stresses empowerment rather than domination.[3] This raises a different story, a divergent narrative; in Augustine's terms—a story about the city of God rather than the city of humankind.

There is another way of considering these questions that emphasizes not so much the differences between religious faith and the academic quest as their similarities. This line of argument was well launched by the philosopher of science, Michael Polanyi, in the 1950s, who demonstrated how both science and religion proceed from a basis of belief, hope, and tradition as well as evidence. Both are thereby forms of personal knowledge (Polanyi 1958). Scientific knowledge is as much tacit, that is framed by preexisting conventions, as any other form. Einstein himself had declared that "as far as the propositions of mathematics refer to reality, they are not certain; and as far as they are certain they do not refer to reality" (Newbigin 1989, 29). They are not fully certain; *but neither are they fully uncertain.*

This simple statement might unlock some doors. A constructionism in full flow argues for the incommensurability of discursive communities, that knowledge is unavoidably local, that mapping cannot reliably occur between domains whether from mathematics to the external world or from one subculture to another. It privileges a model of closure. Consequently, the tendency of postmodern epistemology is to undermine the universalism of the modern tradition entirely, with no remainder. A more judicious argument, following Lesslie Newbigin (1986, 43), suggests that the lamp of the Enlightenment has not been turned out, for the Enlightenment shed real light. Rather, in a century that has seen the abuse of science and technology in advanced weaponry; the ecocatastrophe accompanying unregulated modernization, the unchecked market, and the centralized state; the perversions of the nation-state in fascism and communism; and the evils of Eurocentricity in colonialism; in light of this experience the optimistic, progressive project of the Enlightenment is tarnished. But this does not mean there have not been real and lasting achievements flowing from rational knowledge in science, medicine, and technology, all of which have raised the quality of human life.

Such historical qualifiers concerning the alleged failure of the Enlightenment project invite epistemological qualifiers as well. We need not dismiss all intellectual achievements of white, middle-class males on the basis of their positional attributes alone. The sociologist of knowledge, Karl Mannheim (1952), introduced the concept of relationalism to make the point that the interpretation of events, while relational to an observing subject, was not entirely the product of that person's subject position. There are aspects of that reality about which observers from different vantage points may agree. Indeed, as we share more information of each person's subject position, we may also learn more about the events they are observing. What I am suggesting is that the disabling charge of relativism that is directed against postmodernism may be diverted by an argument that states that the central issue is not relativism but rather relationalism. Relationalism in this respect becomes a synonym for hermeneutics, for the hermeneutic endeavor begins with the inevitable collision between the evidence and the researcher (Gadamer 1986). The interpretive task of hermeneutics provides an alternative and more hopeful solution to the challenges of postmodern criticism than its more radical perspectival alternative (Duncan and Ley 1993).

A radical perspectivalism depicts a closure in social and intellectual life that has been challenged empirically and theoretically. It is predicated on an oversocialized view of human beings that presents people as contained within discursive cages. A passive view of human agency is its consequence. Empirical studies, even those influenced by constructionism, rarely depict such closure. Kay Anderson's (1991) superb interpretation of anti-Chinese racism in Canada, while predicated on the view of racism as socially constructed, recognizes that ideology evolves, is reshaped, and is challenged. Her use of the concept of hegemony is an acknowledgment that even a master discourse is never hermetic; never free from contradiction, subversion, and resistance from human agents. Implicit here is a model not only of learning but also of commensurability, for the marginalized minority knows the intent and interests of those responsible for their minimal entitlement. As bell hooks (1992) reminds us, the racialized understand all too well the character of the racist even though they are worlds apart in their material and social lives. In the same way, claims Deutsche (1991, 29), though David Harvey may marginalize feminism and consider it of no account, "feminism, although it hardly knows everything, knows something about him." There is an area of common understanding beyond the particularities of local knowledge. It is not a matter of closure with no remainder.

A closer examination of the biblical examples cited earlier in defense of perspectivalism would lead to the same conclusion, for though the separation between Jesus and the Pharisees was great, it was not complete. For the Pharisees, resistance to Jesus and his claims was strategic and deliberate as well as socially ordained by their understanding of the law. The high priest's geopolitical strategy and, particularly, his relations with the imperial forces of Rome could not allow Jesus to continue his subversive doctrine. And so, in profoundly ironic terms, Caiaphas addressed the Sanhedrin, the religious council, in words more prophetic than he could ever have imagined: "You know nothing at all! You do not realize that it is better for you that one man die for the people than that the whole nation perish" (John 11:49–50). Power and not simply the limits of local knowledge was at stake. It is that broader and shared human domain that also raises the issue of moral accountability of actions in the Christian tradition and beyond. The Nuremberg (and subsequent) Nazi trials have reinforced the ethical and judicial view that the situational claims of a subculture upon an accused are not sufficient argument to answer crimes "against humanity." On historical, theoretical, and moral grounds, there are not adequate grounds to accept the full constructionist program.

What grounds then remain? It is informative here to take up dialogue with *Logics of Dislocation* (Barnes 1996), which offers an innovative and carefully constructed argument in defense of a strong postmodern relativism in human geography.[4] In his chapter "Rationality and Relativism," Trevor Barnes accepts the view that "the only effective criticism is one that is logically compelling" (1996, 88), and, in the pages that follow, it is through deference to "logical grounds" that he carries out his prosecution of rationality and defense of relativism. This privileging of logic seems to me to beg a number of questions. To what degree do logical grounds become a logical foundationalism? How different is such logic from a Cartesian view of reason? What are the limitations of an argument contained within such bounds? Why is logic seemingly valued above evidence? Is there no place for evidence in such a scenario? Addressing just the last question, the sidelining of inductive methods (Barnes 1996, 90, 106) and the skepticism toward hypothetico-deductive methods does not leave the empiricist with much room to maneuver. But according to my reading, Barnes (like those he criticizes) is not methodologically consistent. In his critique of the work of those following the scientific method, such as Walter Isard or William Bunge, he constantly privileges evidence; the institutional affiliation of economic geographers, accidents of meetings or earlier career experiences, the social climate of the times, are all adduced as explanations of the knowledge they produced. One might proceed further. We have a clear sense that his project is to tell us what really lies behind the production of geographic knowledge that presents itself as innocent. The facts, the knowledge, are interrogated and reveal a world of personalities, social relations, and overarching contexts. But this result does not invite a further deconstruction. We have now arrived at an interpretation, we now see geographical knowledge in not just a different way but a better way. More generally, the relativism that Barnes supports is not only different from but also better than naive realism. Otherwise, why endorse it?

There are at least two issues here. First, the proposition that one account is better than another requires some commensurability between them, grounds for selecting one rather than the other. The grounds for choice in Barnes' discussion, while he claims they have to do with logic, in practice are achieved by evidence (a tendency that is even stronger in his substantive research[5]). Second, in providing a better account, what really happened, the author is abandoning radical perspectivalism. The relativism that he espouses seems to have well-established limits. He defines relativism as "the view that theories are grounded not in broader philosophical systems . . . but in a set of social practices of a given place and time" (Barnes 1996, 82). Are the theories then only a product of those practices, or do they in some way also describe and give an account of a reality? Barnes' own illustrations provide the answer. He examines several cases of geographical knowledge production, and in interpreting them he is claiming that his theories really do illuminate those events. In other words, the concepts that he uses reference more than their own social construction (i.e., the details of his own biography and social practice). In short, his theories provide light, however partial, for the external forms to which they are addressed. If this is true of his work, then so too must it be for the research he criticizes, at least in principle.[6] Otherwise he is guilty of a lack of reflexivity where the relativism ascribed to others is not pervasive enough to affect his own work. Given that the research efforts of both Barnes and those he criticizes produce knowledge that expresses both the external forms of reality and the internal social contexts of their production, then that knowledge is not relative (i.e., merely local) but relational to the subject position of the researcher. The boundedness of the relativism in question directs us not to a radical perspectivalism but to a hermeneutics where the objective is to provide a better interpretation of the formation of knowledge. So, too, despite his apparent adoption of a strong constructionism authored by ethnographic tricksters, Clifford (1986) claims that the new ethnographies are evidence of "better modes of writing."

Other authors have arrived at the same conclusion in their assessment of the sociology of scientific knowledge produced by the constructionist school at Edinburgh, the school that strongly informs *Logics of Dislocation*. In a review of *A Social History of Truth* by the former Edinburgh scholar, Steven Shapin, David Livingstone (1996) observes that "Shapin's narrative is so historically compelling that it seems to refute its own relativist inclinations." Indeed, he repeats Shapin's own first-page observation that "much of this book can be read in the mode of old-fashioned historical realism." Like Barnes, the pursuit of a better account, the marshaling of evidence and not just the deployment of logic, the adoption of a constructionism that is not self-directed, and the acknowledgment that accounts are not socially determined with no remainder; all of this indicates a project that is more properly hermeneutic rather than an instance of radical perspectivalism. Claims to the strong program of constructionism are rhetorical only, and stand betrayed by the work itself. While rhetorically we may be given the choice between competing claims of universal knowledge and local knowledge, such binary thinking does not exhaust the selections available. Just as in the discussion of globalization, geographers and others have rejected on theoretical, empirical, and political grounds the binary distinction of the global and the local (Beauregard 1995; Smith 1995), so epistemologically the choice between universal rationality and local relativism is an incomplete offering.

Writing during the annual ritual of baseball's World Series, this argument might be better understood by a sporting illustration from Walter Truett Anderson (1990, cited in Middleton and Walsh 1995, 31). Three baseball umpires are engaged in some banter as they dispute the finer points of the game:

> "There's balls and there's strikes and I call 'em the way they are" asserts one.
> "There's balls and there's strikes and I call 'em the way I see 'em" counters a second.
> "There's balls and there's strikes and they ain't *nothin'* until I call 'em" insists a third.

The viewpoints of the umpires in this story may be equated with, first, a realist who assumes an unproblematic correspondence between reality and one's own observations, second, what Middleton and Walsh call a perspectival realist, one who recognizes the place of the knower in the construction of reality, and, third, a radical perspectivalist who insists that reality is no more than his/her construction of it. My point is to shift our attention from the third to the second umpire. This middle ground is close to the starting point of a hermeneutics that recognizes that the production of knowledge is accomplished by a researcher who cannot at will erase his or her own humanity and the positionality that it necessarily requires. Nonetheless, shared knowledge and access to external events remain possible. In his exhaustive synthesis of contemporary hermeneutics, Anthony Thiselton (1992, 6) is at pains to rebuff radical perspectivalism while arguing (not unlike Polanyi) for a "commonality of interest shared by multidisciplinary hermeneutics and the Christian community." If we cannot fully escape our ethnocentrism, we can at least attempt to recognize its existence and hope that audiences with different world views will help to make our blind spots apparent to us. This does not lead us to a realist conclusion that, at last, we have achieved the full picture. But it does get us beyond the radical perspectivalism that knowledge is no more than ethnocentric convention. It gets us beyond Nietzsche's fragment, "What then is Truth?" approvingly cited by Rorty (1991, 32) and evoked by Clifford (1986), a fragment in which Nietzsche takes constructionism to its nihilistic conclusion: "Truths are illusions about which one has forgotten that this is what they are. . . . To be truthful means . . . the obligation to lie according to a fixed convention" (cited in Lundin 1993, 37).

What then is truth? This question was first asked not by Nietzsche but by Pontius Pilate. Brought before the Roman governor, Jesus stated "for this reason I was born, and for this I came into the world, to testify to the truth. Everyone on the side of truth listens to me." "What is truth?" was Pilate's sardonic reply, whereupon he turned Jesus over to his executioners (John 18:37–38). Here is Pilate—is it unkind to identify him as a pragmatist?— unwilling to adjudicate between truth claims. For Pilate the suspension of truth permits the implementation of power. But for Jesus, truth is central, indeed the word occurs more than twenty-five times in John's gospel alone. While the Christian might agree that naive realism is not attainable—that to say that we see now through a glass darkly is to acknowledge our finitude, that to claim omniscience is presumption—at the same time, we insist that it is possible to know beyond the norms of our own perspective, our own subculture. We can speak of truth as more than local knowledge.

It is important to note that in holding to a stronger version of truth there are many allies in the academy. Aside from the unacceptably ideological advocates of modernist epistemology, there are more reflexive voices. The critics of Enlightenment certainties do not necessarily revert to relativism, which they recognize as ethically (and politically) inert. Donna Haraway (1996) for one, a prominent voice among feminists, argues for the importance of what she calls situated knowledge in pursuit of "better accounts of the world," against which she contrasts unfavorably the strong constructionist program. For others, the Enlightenment project continues to be upheld as an unachieved objective, though without the triumphalism of the past, by reflective and influential writers like Jürgen Habermas, or within human geography by Andrew Sayer, who claims to locate a middle way between the unacceptable endpoints of naive realism and radical perspectivalism (Sayer 1993). Then there are the serried ranks of hermeneuticians with their hope for plausible, persuasive, though never final, interpretations of events and texts produced by other societies in other times and places. So too, some Marxists, while reproached by the collapse of historical inevitabilities, nonetheless retain a historical sensitivity that rejects ethical as well as epistemological relativism. Should we now seek to understand headhunters rather than change them, asks Terry Eagleton (1991, 385) trenchantly?

In short, in such common resistance to relativism, important parts of the Grand Project, may well not be facing the terminal danger in the Western university that Nicholas Wolterstorff fears. It follows that the proclamation of truth claims need not be quarantined by a pluralism that allows only containment and quietism. One may speak in the public sphere beyond one's own discursive territory, continuing the long-standing method of both science and religion. The apostle Paul, like the contemporary scientist, was a debater, eager to speak wherever there were ears to hear. In this way, truth claims are held up for comparison, judgments are made, ideological commitments are revised or exchanged in light of available evidence and the power of argument. Against the skeptics, we can and do move beyond local knowledge and ethnocentric discourse. Of course the public sphere is never a level playing field, and it becomes important to expose the biases of tradition, local values, and political pressures through reflection and debate that is inclusive in the voices it admits to the forum (Fraser 1991).

There is a final matter to address here that must surely be part of Christian epistemology, and in this there is indeed a departure from the Grand Project. One of the contributions of some postmodern authors has been to encourage a good deal more tentativeness and humility in their conclusions. While we can speak of truth in a plural society, nonetheless our knowledge is typically partial, through a glass darkly, a form of personal knowing, incorporating evidence but also belief, and belief that in certain respects shapes the reception of the evidence. Moreover, the rarified Greek tradition that lies at the origins of the Grand Project has been altogether too cognitive, assuming that its subjects are some cerebral abstraction of whole human beings. When Jesus said, "I am the way, the truth and the life" (John 14:6) he was positing more than intellectual claims. The Hebrew tradition did not recognize the dualism of mind and body, of thought and action. Truth was not simply propositional but was also a matter of conduct, bound up with righteousness and virtue, with "the most excellent way" of selfless *agape* love. This is the vehicle of Christian truth, and it has no legitimate expression in any other form.

In a fascinating convergence with this tradition, the sociology of knowledge is now also throwing up such virtues as trust as central to the transmission and receipt of scientific knowledge. Even more notable in terms of an intellectual convergence is Terry Eagleton's (1990, 413) appeal to the property of love in shaping ethical relationships politically as well as personally. At this point, a Christian apologetic also regards Cartesian rationality as an incomplete mode of understanding and broadens to a much more confessional position that acknowledges the profound insufficiency of cognitive categories before the God who explodes categories in the simple but awesome self-declaration "I AM." What Christians claim to receive is much more than knowledge, a better insight; it is also unmerited grace that exceeds the bounds of human understanding. To put it another way: Christian truth cannot possibly be value free and disinterested. Too much is at stake if this good news is true. Christian truth is propositional, but these propositions reflect a history and require a practice that is impregnated in unmerited love.

Conclusion

This chapter has wandered some way from a particular focus on human geography, an expansion necessary to reveal the broader web of relations that circumscribe any particular discipline. Postmodern criticism reminds us also of the social construction of disciplines as categories, with most of the social sciences defined in the nineteenth century in classificatory acts that were themselves the outcome of the modern mind. In an age that celebrates interdisciplinary border crossings and blurred genres, disciplinary boundaries become ever more conventional and ever less real. To frame possibilities in human geography, then, it is important to address broader intellectual currents, in part because a narrow disciplinary focus leaves too much unsaid, in part because the larger intellectual context has much of relevance to say.

But perhaps current postmodern epistemologies are rather too preoccupied with saying, with linguistic turns, with discursive communities, the printed word, and internet chat groups. Andrew Sayer has correctly observed that among postmodern authors, "Knowledge is only discussed in terms of speaking and writing, never doing" (Sayer 1993, 328). Postmodernism of course is a product of a highly literate society, and one where skeptical reflection in an uncertain world may proscribe action practically, even as it subordinates it theoretically. If the limits of my words are indeed the limits of my world, then we live in an age where ignorance reigns, where uncertainty challenges effective action. One cannot help but note the parallels, but also the discontinuities, with the opening verses of John's gospel: "In the beginning was the Word." Is this another case, as Milbank has taught us to expect, of social thought colonizing and appropriating earlier Christian themes? But consider also what has been lost in the transfer from the Word to words. Jesus, the Word, is not just unstable discourse but an active, redemptive, historic figure. He is, writes John, the light of men and women. "The light shines in the darkness, but the darkness has not understood it" (John 1:5). Here is an enlightenment that shines through the cynicism that seems to be part and parcel of postmodern word games, an ontology of love that challenges the ontology of violence in the dark world of Nietzsche and his intellectual offspring. Here is a model of power as servanthood that empowers others in contrast to the pervasive postmodern disclosure of power as domination. "The Word became flesh and lived for a while among us. We have seen his glory, the glory of the one and only Son, who came from the Father, full of grace and truth" (John 1:14). Whose (W)ord? Which (E)nlightenment? In opposition to postmodern relativism, the Christian faith directs the searcher to the creative and loving Word, full of grace *and truth* . . .

Notes

[1] While absolving them of the idiosyncrasies that remain in the text, I am extremely grateful to colleagues and friends who have helped improve this chapter, particularly Trevor Barnes, David Livingstone, and Iain Wallace. With others I am also very thankful to Henk Aay who labored hard in the completion of this book and in arranging the conference at Calvin College that gave rise to it.

[2] The literature in this field is now legion. Within the humanistic tradition, I offered applications of a constructionism influenced by social phenomenology in several papers in the late 1970s, and made a fuller use of the constructionist framework of Berger and Luckmann (1996) in Ley (1983). Work of considerable maturity that follows a constructionist position has been published by geographers in such fields as ethnic and racial studies (Anderson 1991), feminism, including the seminal if controversial contribution of Rose (1993), the new cultural geography (Barnes and Duncan 1992), postcolonialism (Jacobs 1996), and most recently in economic geography (Barnes 1996). A panoramic vista of a good deal of this work is painted by Gregory (1994).

[3] This is not to claim for a moment that the historical record is one that shows the church (or for that matter individual believers) consistently living up to the high calling of their founder. However, rarely has that history been reported evenhandedly by secular writers. Much secular history of the church and mission has included ideological baggage that has drawn it consistently away from a balanced interpretation of the Christian presence in the world.

[4] The reader should know that Trevor Barnes is both a colleague and a friend. His book is discussed because it is the fullest and most knowledgeable defense of postmodern epistemologies (a term that he might find to be an oxymoron in geography). Moreover, I know his arguments fairly well, for I am here continuing publicly a number of private conversations we have enjoyed in recent years. Rather cheekily, I might add that although we sometimes choose to disagree, we have had little difficulty understanding each other.

[5] Not only is Barnes' own empirical work conventional in method in light of the elaborate critical structure he raises, but so too is the work he presents as exemplary, including, for example, Massey's interpretation of economic restructuring in South Wales (Barnes 1996, 98). The point is not to object to Massey's work, but to draw attention to its unremarkable methodology. The major requirement seems to be the unexceptional property that good work should be comprehensive in its synthetic ambitions, aiming to include "the broader geographical context in which acts are set" (with which I fully concur). Note too the implicit assumption that Massey is telling us something about the "reality" of South Wales. Good work clearly escapes the iron cage of its discursive context, enough at least to present us with knowledge that is relational

rather than relativist, i.e., it escapes the net of radical constructionism (but see note 6).

[6] In fact Barnes is inconsistent here, for in his examination of four spatial scientists, he claims that "the theories and models that each advocates reflect *only* the local context—in which they were propounded" (1996, 105, my emphasis). This is to evoke the strong program of radical constructionism. But it begs the question of why the accounts presented by Barnes himself or his exemplars, like Massey (see note 5), offer "better knowledge" and are not equally the product of a totalizing social context; unless radical perspectivalism is itself unequal and therefore localized in its incidence! Reflexivity could not permit that possibility, for "local practice always comes first" (113).

Chapter 3

A Christian Reading of the Global Economy

Iain Wallace

THIS CHAPTER EXPLORES some of the challenges that Christian geographers face in coming to grips with the global economy. My orientation and choice of issues are governed by those dimensions of the world around us traditionally subsumed within "economic geography," although discussion is not confined to the conventional systematic interests of the subdiscipline. Rather, I affirm the seamlessness of human experience, the embeddedness of social life in the natural environment, and hence the authenticity of concepts of geography as a radically integrative discipline.

My starting point is one shared by all students of the contemporary world: How are we to understand the dynamics of change in a time of rapid and differentiated economic transformation? To tackle this question forces geographers to reflect on how their conceptualization of the world is itself made problematic by changed empirical realities. Those who live within an established interpretive tradition, such as biblical Christianity, are further challenged to reassess the possibilities of sustaining that worldview, in the light not only of these global transformations but also of the postmodern rejection of metanarratives. Their reflection on the consequences of new ways of seeing the world may then achieve a measure of integrity by promoting faithful responses to the novel issues that are thrown up by transformations in the global economy and its sociocultural matrix. Of course, these steps are not discrete and sequential: They are blurred by the reciprocal and ambiguous interweaving of fact and theory; of faith and reason; and of belief and practice. Nor can an agenda of this scope be reviewed other than summarily in the present volume. But I take courage from Douglas Hall's suggestion that the risk of simple-mindedness inherent in trying to knit together multiple fragmented knowledges may be the cost that contemporary Christian academics have to accept as "fools for Christ" (Hall 1991, 200).

World(s) in Transition

We live in a world and a time of profound transformation (Johnston, Taylor, and Watts 1995). The geopolitical and geo-economic order has shifted dramatically within the past generation. The stability of rival Cold War hegemonies has collapsed. As one consequence, the world economy of capitalism is now all but universal, and its penetration of society is promoted by transnational corporations and multilateral institutions (such as the World Bank and International Monetary Fund) alike. Armed conflict, and the threat of more, has not died away, but international rivalry is contested more in the economic than in the military sphere. The United States stands alone as a superpower, although its hegemony has been significantly trimmed (Agnew and Corbridge 1995). The regionalization of the globe into a structure anchored by North America, Europe, and Japan and defined by flows of trade, finance, and cultural dominance captures the basic geography of the new order (Stallings 1995). This simplification of the world map reflects also the powerful forces of technological change that have made possible the "space-time compression" of the capitalist world economy (Harvey 1989).

But the ideological triumph of liberal capitalism has not, contrary to Fukuyama (1992), brought history to an end. Within most capitalist states, the legitimacy of the institutions that delivered prosperity in the post-1945 era is severely challenged. Government, and the broad public sector (including education), is held in low esteem, having lost (forfeited?) its moral authority as an agent of a widely recognized public good. The fiscal crisis of the state, in an era of intensified global competition and a culture of endlessly expanding "entitlements," is sapping the public will and capacity to maintain redistributive mechanisms that increase the security of the vulnerable and promote social cohesion. A private sector increasingly seen to be driven by a narrowly defined profitability, slashing employment while raising executive compensation, is adding to social and personal dislocation. The values of those who command the "good jobs" of the information age are rarely shaped by a fully inclusive vision of civil society.

In many other parts of the world, the liberal capitalist hegemony is equally under fire. The imposition of structural adjustment programs on indebted governments in Africa and Latin America has inflicted the blinkered orthodoxy of the international financial institutions on civil society, gutting social spending and fostering increased dependence on export monoproduction at considerable environmental cost. After fifty years of Western-style "development," the verdict in many societies is that not only has it failed, but that it is a Trojan horse. Increasingly, the Enlightenment worldview and secular values of Euro-American societies are being challenged and rejected by erstwhile subaltern peoples. The Iranian revolution; the revival of Islamic fundamentalism from Algeria to Afghanistan; the active resistance of indigenous peoples (in places such as James Bay, Chiapas, and the Narmarda Valley) to the social and ecological destructiveness of "development-as-progress;" and the urbane rebuke by Asian political leaders of the corrosive individualism of Western democracies all point to unresolved problems within the worldview of quasi-hegemonic capitalism (Kothari and Parajuli 1993; Reid 1995).

Meanwhile, the social and environmental consequences of a steadily expanding "human loading" (population *times* per capita consumption) of the planet are becoming clearer, albeit fitfully. The neomalthusian concerns of the 1970s have waned, but the social and political implications of continued population expansion in poor and environmentally degraded nations have prompted renewed warnings of major dislocations in the offing (Homer-Dixon

1993). International recognition that disruptive, anthropogenically induced climatic change is afoot has largely been secured in the scientific community, but its implications are politically divisive within and between nations, and commitments to responsive action are, with few exceptions, weak and localized. Rapid economic growth in southeast Asia, and particularly China, promises rapid increases in the human loading of the biosphere, but there are few signs of willingness on the part of the rich industrialized nations to moderate their intensity of resource use and waste generation to free up any of the disproportionate share of global environmental carrying capacity that they have appropriated.

The employment implications of global population growth intersect in complex and politically charged ways with other processes that are transforming the world. The depth and pervasiveness of the revolutionary changes associated with the emergence of computerized, information-rich societies is only gradually becoming apparent. The impact on humanity's mental activity is likely to be as great as was that of the Industrial Revolution on physical activity, with comparable social upheavals and new challenges. What does seem to be clear is that within industrialized capitalist economies the transformation of work is marked by considerably greater polarization of earnings than was the case under post-1945 (welfare-state) fordism. A rising underlying rate of unemployment and a bifurcation of the employed workforce (including the special circumstances of the former state-socialist countries of Europe) creates social stresses that are magnified by a growing volume of industrial employment and output in a specific group of lower-wage economies, the "newly industrializing countries." The benefits and liabilities of a global labor market are not evenly distributed socially, geographically, or intergenerationally at any scale of analysis. Moreover, they are fundamentally marked by differences of sex, ethnicity, and other personal identifiers (Noble 1995, Wood 1994).

Geographical Interpretations of Global Change

The past decade has been an exciting time for geographers to engage with the challenges of global change. There has been a constant stream of substantive issues (outlined above) to document and analyze; the intellectual freedom to explore a wide range of philosophical and conceptual approaches to enquiry has broadened; and the growing integration of geographical discourse into the whole arena of debate in the social sciences, cultural studies, and relevant biophysical sciences has brought welcome academic synergy and much personal satisfaction to those engaged (Gregory, Martin, and Smith 1994). This ferment has been reflected in the appearance of textbooks that provide a fresh approach to human geography as a discipline that is able to provide an integrated but differentiated analysis of global change (Johnston, Taylor, and Watts 1995). Closer to the core of conventional economic geography, general texts that situate the geography of the capitalist world economy within a broad political-economy perspective (Knox and Agnew 1994, Wallace 1990), and more focused studies of particular themes (e.g., Corbridge, Martin, and Thrift 1994), have widened the intellectual horizons of students in an often conservative field (at least in North America). Moreover, the globalization of the world and the convergence of erstwhile compartmentalized knowledges has created an environment in which geographers have found productive, albeit provisional, resolutions to some of their traditional dilemmas. Linkages between different scales (global/local, national/international/subnational) and between different sensibilities (place/space, the sources of "otherness") are more easily articulated within the current climate of open and fluid engagement with disparate conceptual frameworks and the varied phenomena in which they find significance. All this, of course, is one way of acknowledging that postmodernity has genuine appeal.

Yet in his creative and penetrating account of how economic geographers have construed their intellectual task over the past thirty years, Trevor Barnes argues that the break with a positivistic spatial science has been less complete than might appear (Barnes 1996). His four exemplars of alternative approaches (Harvey's Marxism, Sayer's critical realism, the British "localities" studies, and the "flexible production" debate within industrial geography) all bear marks of significant, though by no means unambiguous, continuity with Enlightenment thinking. Specifically, he finds evidence of: "the continuing sense that progress is possible . . . ; conceptualization of the subject [as] 'Enlightenment man' . . . self-directed, autonomous, and rational . . . ; a tendency toward totalization . . . the impression that there is always only one story to tell . . . ; [and] the overwhelming sense that our theories are [reflections] of some real set of economic events" (Barnes 1996, 40-41).

In contrast, his three exemplars of an emergent "'post'-prefixed economic geography" (Graham on overdetermination, "postcolonial" geographies of development, and feminist work on labor markets) provide evidence of (1) accounts that recognize their own partiality and provisionality; (2) the shunning of single causes in favor of embracing "the ineluctability of difference"; (3) acceptance that "the supposed Western universals of rationality and progress [are] not universals" but ideological and regionalized constructions; (4) accounts that recognize the power relations embedded in their discourse; and (5) acceptance of "the centrality of political practice" (Barnes 1996, 42). Barnes himself examines the work of Sraffa, Harold Innis, and Lukermann to find forerunners of a "different kind of economic geography," one that is congruent with a suitably modest assessment of what geographers can deliver. His "dislocating" analysis of concepts that have shaped recent geographical research allows him to commend an approach that is marked by the "shunn[ing of] closure, universals, and dogmatism and [the] embrac[ing of] openness, context, and reflexivity" (Barnes 1996, 251). Martin offers a comparable specification for a renewed economic geography, one that is "much more multidimensional, multiperspectival and multivocal" (Martin 1994, 45).

The principal tension that Barnes uncovers is thus between "modernist," totalizing and "progress"-ive accounts and the fractured and relativized explanations that are the hallmark of postmodernity. His focus is epistemological, dissecting modes of creating geographical knowledge rather than ontological. But in exposing and questioning the hidden assumptions that his exemplar authors display, he prepares the ground for an engagement with the contending worldviews that shape their work.

Barnes legitimately casts David Harvey as a deeply committed exponent of the emancipatory and universalist goals of "the Enlightenment project." *The Condition of Postmodernity*, in particular, provides an insistently, though ambiguously nuanced, Marxian analysis of capitalist dynamics that charts the interpenetration of contemporary cultural, political, and economic transitions (Harvey 1989). This worldview remains committed to energizing an ethical, historical project for humanity, releasing men and women from the alienations and privations of the now dominant social and economic order. Harvey formally acknowledges that the dialectical relations that constitute the essence of his metanarrative necessitate respect for "the particularities of difference" that are constantly modulating the outworking of the universals of capitalist relations. But his totalizing concept of capitalism undermines his capacity to validate situated knowledges that are not wholly derived from relations within the mode of production, such as differentially gendered experiences (Massey 1991).

Harvey struggles more explicitly with the tension of justifying a universal ethic in a world of situated, and therefore *relative*, knowledges in his account of a tragic industrial "accident" in the North Carolina "Broiler Belt" (Harvey 1993). He grants "the seriousness of the radical intent of post-structuralism to 'do justice' in a world of infinite heterogeneity and open-endedness," but he fears that at the end of the road taken by acknowledging "contextualized justices" is "a void or . . . a rather ugly world" in which the morally repugnant are allowed to celebrate their own particular form of "difference" (Harvey 1993, 103). Moreover, Harvey is concerned that poststructuralist ethics tend to be conceived at too localized a scale to seriously threaten the hegemonic power and practices of capitalism. For American consumers of franchised fast food to respond effectively to the exploitative and callous employment conditions of broiler processing in a southern, single-industry town requires a concept of justice capable of being made effective at a distance. This implies a systemic politics. It may also imply that a particularized identity, often forged in the workplace, is recognized as the creation of alienating social relations that a generalizable expression of justice demands be challenged, even if painfully for those whose self-definition is threatened. Harvey finds his bearings for making these judgments in "a modernized version of historical and geographical materialism, which . . . [s]truggles to bring a particular kind of discourse about justice into a hegemonic position" (Harvey 1993, 115). Yet partisans of a concept of justice that reflects the experience of the oppressed necessarily privilege their own sense of "otherness" over against that of other "others." If, as Harvey claims, the universal ethic cannot "be imposed hierarchically from above" but must be locally negotiated, a tension arises that makes the pursuit of progressive politics more problematic than he has been given to acknowledge hitherto (Harvey 1993, 116-17).

But there are no easy answers for critical geographers of different ideological persuasions either. In their recent interpretative reworking of the field of international political economy, Agnew and Corbridge find it difficult to articulate a convincing "Big Story" (Agnew and Corbridge 1995). They seek to encourage the breakdown of the "established representations of space" associated with the coercive global hierarchies arising out of state-system rivalry, on the one hand, and the unregulated operations of the market, on the other. They claim to be working toward the creation of "new representational spaces" flowing from a worldview that envisions a radical diffusion of power and a real reciprocity of interaction between peoples. For them, a set of universal principles consistent with the empowerment of local identities would draw on "the diverse discourses of market socialism, internationalism-multilateralism, and a series of more local oppositional imaginations" (Agnew and Corbridge 1995, 211-12). They see in the evolution of a novel *global* consciousness among diverse subordinated groups the prospect for an ethic of care to develop that tangibly subverts an exclusionary "distance-decay model of morality." They argue that Rawls' theory of justice can be applied to the international arena, where it is capable of providing a set of universal principles that respects significant difference and justifies the preferential standing of disempowered others (Rawls 1971). But they also note that they "have little sympathy" with the postmodern sensibilities of others who refuse, by inference amorally, to identify with at least some minimal claims of Rawls' rationalism and universalism (Agnew and Corbridge 1995, 214-17).

Yet to recognize a totalizing discursive colonialism in Rawl's worldview can be both a principled and geographically informed position (Wallace and Knight 1996). Significantly, critiques precisely of the rationalism and implicit, if not explicit, universalism of postwar economic-development theory have come to the fore recently. Mehmet argues that the Eurocentric bias of mainstream development theory is notably evident in the enshrinement of individualistic Western-style rational market behavior as a universal norm (contra the evidence), and in the refusal to validate the ethical imperatives of other cultures (Mehmet 1995). Even more directly, the dismissal of the spiritual dimension of humanity and of its development goals within a thoroughly secularized discourse is being identified as a major source of the "failure" of development programs and of the resistance of non-Western peoples to the agenda of "progress" (Ryan 1995).

Issues Facing a Christian Worldview

Christian geographers engaging these analyses of global economic change face a number of challenges. The most searching is our response to the demands of Barnes' "'post'-prefixed economic geography." His objective in *Logics of Dislocation* is to liberate the field from the constricting grip of essentialism: from overarching explanatory discourses that impose order and meaning on a fractious reality, suffocating the contextual vitality and otherness of objectified peoples. But this postmodern critique of universalizing metanarratives has profound implications for Christians. We are a people of the "big story" and are accustomed to proclaiming it as God's way for the whole earth. Both theologically and pastorally, the church through the ages has had great difficulty in grappling with the claims of different others, within and beyond its ranks. Catholicism and Protestantism alike have provided fertile ground for institutions and parties claiming to be sole incarnation and interpreter of God's will, attempting to shape the world in their own images, and being quick to condemn those who are differently persuaded. Moreover, until comparatively recently, the history of the worldwide church has been intimately associated with the worldview and political economy of Western societies. Insofar as it is the claims of Enlightenment thinking that are now being contested, within geographical discourse as more widely, they are the tradition that has most shaped Christian thought and practice. So perhaps the first constructive response of Christian geographers to Barnes' work is to recognize and listen to the dislocating voices that have begun to emerge within the contemporary church.

The early exponents of Latin American liberation theology have recorded how they had to deconstruct the entrenched (because it had been so completely and unselfconsciously assimilated) worldview of European academic theologians in order to create space for their own understanding of, and experience of, the kingdom of God (Bonino 1975). They struggled to the recognition that idealist theological abstractions were a perversion of the gospel that needed to be spoken contextually to their compatriots. The poverty of subject peoples and collusion in their disempowerment had been, and was still, too complacently sanctified by the church of the *conquistadores* and their successors. Of course Marxism provided dislocating hermeneutic assistance for many writers, but this was harnessed to an authentic prophetic tradition in the Scriptures calling God's people to a love that issued in justice and the shared enjoyment of material well being.

I want to suggest that a comparable process is under way today as the voice of Christian aboriginal peoples is being addressed to mainstream theologians. One of their avenues of engagement is to radically question the worldview that shapes Western environmental thinking, revealing that a society built on the unbridled exploitation of nature has been too readily baptized by the churches at the core of the capitalist world economy. Consider, for instance, the argument of George Tinker, a Cherokee theologian (Tinker 1996). He dislocates the taken-for-granted prioritization within Euro/American theology of the temporal over the spatial—something that geographers have long decried in other contexts! Attachment to the land embodies "the primary metaphor of existence for Native Americans." For them, the first question raised by the concept of *basileia tou theou* (the kingdom of God) is not the when but the where. Tinker notes that this metaphor is not used in the Old Testament, which is notable given the central and paradigmic significance of the land in Israel's self-understanding as the people of God (Wright 1983), but that the image of the kingship of God is invariably associated with God's acts in creation. "If the metaphor has to do with God's hegemony, where else is God actually to reign if not in the entirety of the place that God has created?" This insight undergirds the native spirituality that considers the achievement of harmony and balance within all of creation as a primary goal. "[N]o one can be left out of the *basileia*," and the call to repentance because "the *basileia* is at hand" (Mark 1:15) entails the return to a proper relationship with both the Creator and the rest of creation (Tinker 1996, 126). Acknowledging this framing of God's demands poses awkward questions to those interpretations of the dominion motif in Genesis by which voices in the Western church have justified an expansionary and intensifying global economy.

Christians would be wise not to subvert these challenges prematurely by the orthodox theological responses that God's kingdom is not confined to the earthly creation; and that the outworking of salvation does, indeed, have a history. Douglas Hall warns of the Western church's track record of reacting defensively to the Holy Spirit's capacity to "reintroduce life and nuance and movement" through peoples and in contexts that dislocate its theological paradigms (Hall 1991, 104). Beyond its substantive focus, Tinker's work is representative of an increasing variety of Christian theologies by members of the people of God for whom the European cultural tradition is not their own. However much many Euro/American Christians have sought to grow in recent years in respect for the culturally other, both within and without the church, I sense that many of us still find it novel to encounter well-developed theologies that display faithfulness to the biblical revelation but also to unfamiliar basic assumptions.

One of the most challenging of these assumptions has to do with the *theological* understanding of non-Western people's premissionary history. Numerous indigenous beliefs and practices that were invariably and undiscriminatingly appraised as heathen error by "Christ-bearing colonizers" (however well meaning and faithful to their own sense of discipleship) have been reassessed through very different eyes by the churches of the Third and Fourth Worlds (Baldridge 1996). The too-ready rejection of the spirituality of native peoples by those missionaries who saw in it a dangerous syncretism, or worse, is being confronted by others now able to reflect theologically on the cultural beams in the eyes of the Western church. Nor is it only with respect to the natural environment that the values of Western Christianity are on trial within the global community of biblical faith: The discursively privileged values embedded in social relations, political institutions, and the organization of economic activity among peoples of Euro/American origin are being dislocated by lived embodiments of difference that are no less faithful to God's revelation. This is a profoundly discomforting—but potentially enormously liberating—development! It gives Christian geographers both epistemological and substantive resources for engaging with equal honesty and seriousness the task that some of our leading secular colleagues have embarked upon.

Our task is not an easy one. However great our sympathy with the critical iconoclasm of postmodernity, we cannot renounce our scandalous commitment to the one who is "the way, the truth and the life" (John 14:6). Having been embraced by the love of God and incorporated into the community of the Church, we know that there is a foundational "big story," linking God redemptively to creation and the human race. We are committed to its telling, faithfully and imaginatively, not least with a sensitivity to the context of narration that is informed by our geographical understanding. We need to acknowledge that our particular cultural tradition shapes and selectively filters the story, and hence be alive to the provisionality and partiality of our personal (and our churches') discourse. The Western theological tradition (within which the contributors to this volume stand) has been thoroughly modernist in its unreflexive presumption of universality, imagining that the ontological omnipresence of its divine subject guaranteed the globally invariant adequacy of its epistemological strategies and the truth of its doctrinal formulations. Bearers of this tradition will therefore engage their own narrative critically, and can most appropriately do so by testing its adequacy against that of other Christians who see the world from a different place. Yes, Christians do have a "big Story," but there is no single definitive recounting of it. The power of the tale lies not in our always inadequate forms of metanarrative, but in the person of the Author, who continues to engage human beings contextually, where they are. "Only God can have 'the whole world in his hands'; human beings can grasp the whole only by taking hold of that part nearest them" (Hall 1991, 365).

Christian geographers, therefore, have no cause to be reluctant in embracing the "'post'-prefixed" agenda of Trevor Barnes. (They might, however, graciously correct his last sentence [Barnes 1996, 252]. Forgiveness, not perfection, is the divine characteristic normally contrasted to human erring, one that significantly dislocates conventional images of God). That Christians in other places than the Westernized hearth of the church and its settler colonies are increasingly dislocating the dominant theological discourse is entirely to be welcomed. The practice deserves to be emulated by us all, insofar as it involves a perceptively critical reading of the specific place and time *(kairos)* out of which our theologies arise. This will involve potentially uncomfortable interrogations of the metanarrative, or "big story," that we, as individuals and faith communities, express. It is likely, for instance, that the truth to which we subscribe will be framed less in the language of proposition and comprehensive finality and more in that of networks of relationships to the Living Word, in whom all things cohere (Col. 1:17) and who is constantly able to do new things that refuse to be confined to our categories. A reflective reevaluation of the history of Western missionary enterprise will also be in order: One that can discern its good and godly thrusts as well as those moments that have deservedly given rise to a guilty conscience (Verstraelen 1995).

What, Then, Shall We Do?

Douglas Hall's deeply challenging, book-length call to do theology contextually provides a charter for Christian geographers (Hall 1991, especially 122-26). He identifies quite explicitly the necessity of a theological "place-consciousness" that is attuned to the localization of human reflection within a global arena. He recognizes that within the church, no less than within other human organizations, local commitments that lead to neighborly praxis are usually the best guarantee of effective engagement with the needs of the world. In addition to the broad epistemological ramifications of this charter, which have been explored briefly above, one can identify avenues of response that more immediately engage the subdiscipline of economic geography.

The state of the contemporary world that was sketched at the start of this chapter could easily induce paralysis, even within the church. So many issues in so many places cry out for attention, yet globalization appears invariably to be associated with a loss of points of personal or institutional leverage. Moreover, in the field of social ethics, Christians have discerned different and often polarized normative foundations for action, leading to fragmented and often contradictory praxis. Hence, for instance, the divergent readings of the nature of capitalism (in terms of the balance it permits between freedom and justice, economic growth and equity) and of the scope and legitimacy of the state (Gray 1991). It is consistent with the argument of the previous section that one would expect to find local differences in the priorities of the Christian community and in the forms of action to which they give rise, even amongst relatively similar cultures. For example, the differing relative weight given within Catholic and Protestant social philosophies to organic versus individualistic concepts of society underlie significant contrasts between the European Union and the United States in terms of attitudes toward regional disparities, modes of healthcare provision, and the scale of support to the unemployed. But rather than simply celebrating such differences, Christians are called to allow expressions of discipleship in other places to engage critically the necessarily provisional and imperfect modes of obedience developed in their own context. In this way, the dynamics of globalization/localization can be made to serve an ethical function, coaxing the universal church to approximate in its faith and practice those expressions of truth into which the Holy Spirit works to collectively guide it (John 16:13).

In the current neoliberal international order, in which the power of global markets and corporations has eclipsed that of the institutions of territorialized civil society, effective (as well as contextually faithful) action is often most readily identified locally. Herman Daly and John Cobb have, for example, provided creative guidelines for responding to the ecological and socio-economic ills of the United States (Daly and Cobb 1989). Yet problems with their carefully argued and substantial effort illustrate the radical challenges facing Christian visionaries of a contextually authentic and faithful society. Their "policies for community in the United States," for example, seem to reflect two weaknesses. One is the seeming inadequacy of the community institutions at various scales that they advocate to command sufficient power and loyalty to outflank transnational capitalist enterprise promoting a culture of consumerism. This weakness invites the more robust argument of David Harvey's Marxian critique (Harvey 1989). Another is their insufficient recognition of the localization of their *theological* framework in time and space. Drawing on other examples of recent social reflection in the American church, Hall questions whether a vision of community grounded in "the biblical and republican traditions that the small town once embodied" takes sufficiently seriously the demise of modernity and American Protestantism's history as a cultural bulwark of the illusions of modernity (Hall 1991, 41, quoting Bellah 1985). The quasi-autarkic community that Daly and Cobb appeal to reflects the "nostalgia for the same brand of idealism and human potentiality that is the very stuff out of which the modern vision was constructed" (Hall 1991, 41). The consciousness of failure and suffering that is more readily found in theology created out of European experience would serve as a salutary foil. (And the economic geography of Europe would provide a less sanguine reading of the praxis of autarky).

Notwithstanding, Daly and Cobb offer much pertinent material that deserves to be more widely taken up theoretically by geographers. For instance, their basic recognition that a world economy embedded in the biosphere cannot grow indefinitely, no matter how much it becomes "post-material," should direct our attention more actively to such concepts as the "ecological footprint" of societies at all spatial scales (Wackernagel and Rees 1995). To some extent, this is being achieved through such means as the study of agro-industrial commodity chains (Friedland 1994); but there is enormous scope for further analysis of the intricate global web of environmental and sociopolitical interdependencies that sustain modern economies. This will require the elaboration of a biblical ontology of human activity in the biosphere, one that is more contextualized and a truer reading of the human condition than that embedded in prevailing discourses of economic geography. Western theological reflection on the implications of the earth's finitude has been gradually developing since the 1970s in tandem with the growth of the modern environmental movement. The title of John Taylor's book, *Enough Is Enough*, expresses a basic ethical challenge to those nations and individuals who constitute the world's materially affluent minority (Taylor 1975). A number of authors have expounded the Sabbath and jubilee ordinances of the Mosaic covenant as the grounding for an ethic of moderation (Wright 1983). In addition to the ecological-stewardship and social-justice strands of these regulations, Sharon Levy, has identified a relational one: that Yahweh set bounds on "His" own creative ability "so as not to overwhelm creation" (Levy 1995, 87). In resting on the Sabbath, God exercised a capacity for self-control that humanity is called to emulate. Yet, patterns of living that decisively dislocate the priorities of the prevailing secular culture appear only as oases within the church and beyond its bounds. Critiques of growth-obsessed economic systems tend to attract no more of a sustained response than did the original "voice in the wilderness."

If progress in embedding the biophysical constitution of the earth as a foundational element in geographical theorizing about the world economy has been disappointing, so too has that in securing a subject ontologically different from "Enlightenment man." There are dangers of essentialism in pursuing this claim, but it is incontestable that models of a more diversified and socialized human identity would lead to significantly different theoretical outcomes.

Accommodation of gendered difference within the hegemonic metanarrative of mainstream economics continues to be resisted and remains absent from most undergraduate economic geography textbooks (Waring 1988). Moreover, as noted above, validation of non-Western, nonmarket values receives, at best, only lip service in models of economic development. Western Christian reflection in these areas has been limited until recently, not least because of the domestic difficulties that the church has had in coming to terms with valuing difference. Considerable progress has, however, been made with respect to valuing women and deconstructing a culture of patriarchy, and this gives hope that comparable advances may be possible in other fields (Van Leeuwen 1993). It certainly reinforces the argument for Western Christians to engage seriously and contextually with theologies emanating from the church as it is rooted in other cultures.

The interpenetration of the local and the global, which is one of the distinctive hallmarks of postmodernity, provides Christian geographers with an unprecedented challenge and opportunity. A church that begins the process of dislocating its prevailing theological worldview in response to an enlarged sense of God's purposes will begin to implement its new vision where it is, building on a deeper awareness of the place wherein it is set. But it will soon be faced with its responsibility and potential as a global actor. In the past, Western Christians have fulfilled their theological understanding of mission expressed in universalistic models—a Constantinian papacy, a bureaucratic World Council of Churches, or the cultural imperialism of American-style satellite tele-evangelism. An adequate theology of the local-global nexus demands, however, a new openness to difference. Faithful responses to God's call to the church will involve complementary forms of obedience derived from the contextualized experience of Christians in a world of places. Geographical, as well as theological, sensibility is a necessary basis for this. The triumph of the market is currently a global reality, but in a world of those who do not live by bread alone, there is an adequate basis of hope to work to dislocate its discourse and its outcomes.

Chapter 4

Christian Worldview and Geography: Positivism, Covenantal Relations, and the Importance of Place

Janel M. Curry-Roper

THE DISCIPLINE OF geography, like all others, is situated within the context of the general worldviews that have dominated historic eras. Presently it sits between an Enlightenment-positivist worldview, grounded in a search for universal laws, and a worldview increasingly dominated by postmodern thought and its emphasis on context and locale. The emphasis on context is especially evident in the resurgence of interest in "place" within the discipline.

The tensions between adherents of these two perspectives, characterized by the contrasting concepts of space and place, or ideographic and nomothetic approaches, are not new. The debate, however, has been given new life with present postmodern challenges to the dominance of the positivist perspective. In this chapter, I attempt to describe both the positivist worldview and the postmodern worldview. Insights from the place literature, drawn from across disciplines, are compared to the thought of two prominent critics of the positivist school in geography from a previous era, Carl Sauer and Richard Hartshorne. Both have become the focus of contemporary retrospectives in light of present discussions. Finally, I attempt to show the relationship between certain aspects of the place literature and a Reformed, neo-Calvinist understanding of society and its relationship to nature. In doing so, I hope to show how Reformed Christian scholars have shown an alternative to the polarization of the universalizing orientation of positivism, and the relativizing and localized perspective of postmodernism.

The Worldview of Positivism

The dominant paradigm for the sciences and social sciences, from which geography draws, is rooted in Enlightenment thought, and is still best characterized by the methods and assumptions of positivism. The Enlightenment assumptions of rational objectivity and the reducibility of phenomena, including society, to the level of the individual (reductionism), continue to dominate geography. This approach to knowledge, which has historically accompanied the production of scientific facts, is Cartesian reductionism. It is characterized by the practice of breaking a problem down into discrete components, analyzing these separate parts in isolation from each other, and then reconstructing the system from the interpretations of the parts. For example, descriptions and analyses of human responses to environmental or economic change have typically centered on individual responses and actions. Regional or global analysis has likewise assumed that large-scale trends are merely the aggregate of individual responses to change. Scholars with this approach have not been interested in a complete understanding of a specifically situated phenomenon but in partial understandings of widely dispersed but similar phenomena.

Rationalism is another aspect of this present paradigm. Underlying it is the assumption that more objective information will bring to light the rational answer on which all can agree. Once the correct application of individual reasoning (or the methodological apparatus of science) is applied, rational interests are assumed to converge. The result is the splitting of the scientific from the moral, and the objective from the subjective. Positivism and the methods of science have attempted to completely eradicate the subjective; thus seeking to prove that science must necessarily be objective. The scientific method, based on observation and experimentation, becomes the only means for acquiring reliable knowledge (Botha 1988, 35-36).

The spatial-analysis and physical-geography traditions have been largely formed by a positivist perspective. Physical geography, close to the natural sciences, began to be influenced by positivism and its more systematic emphasis by the late nineteenth century (Schaefer 1953, 229). A more recent emphasis on the scientific method and reductionism came with the quantitative revolution and the growth of the spatial-analysis tradition. Those working within the spatial-science tradition, in economic geography in particular, accepted methodological individualism and its *homo economicus* model of human nature that ostensibly provided a framework for analysis that could be applied in all places at all times. Communal bonds, if existent, are viewed as a backdrop against which individuals determine their self-interests (Miller 1992, 23, 28-29).

These positivistic spatial perspectives see the geographical world of space and particular places as simply the locations of spatial attributes. Scholars with this perspective tend to use quantification in order to find generalizations that apply across space that can simplify human experiences in order to isolate those aspects deemed most essential (Sack 1989, 144).

Fred Schaefer, a logical positivist and one of the first strong advocates for positivist perspectives in geography, limited the field of geography to the formulation of laws governing spatial distributions, excluding individual facts or phenomena. He emphasized causal relationships and the reduction of the relevant factors to a limited number of phenomena and also championed the goal of developing laws applicable on a worldwide scale. He saw exceptionalism in geography, the description of unique occurrences, as a backward perspective that kept the discipline from attaining the status of other sciences (Schaefer 1953, 227, 229-30).

I argue that the result of positivist perspectives in geography and other disciplines has been the obliteration of a rich sense of place and of our relationships with others. Differences among places have become characterized as the result of individual actions that respond to measurable forces such as distance, the presence or absence of particular resources, and larger economic forces, rather than arising out of a long history of social embeddedness in a place and with others residing in the place. Scientific knowledge has been limited to that knowledge that transcends place and is universally applicable to the exclusion of local knowledge that is based on the "thick" description of complicated interrelationships among entities. The result has been a movement in a direction of limiting our "imagination" in the process of understanding geographic systems.

Geography and Place: Universality or Uniqueness

The field of geography, somewhat uniquely, has always struggled with the tensions of space verses place; the search for universals verses the recognition of the unique; a nomothetic approach verses an ideographic perspective; the objectivity of science verses the ambiguity of lived experience; structure verses agency (Entrikin 1991).

The rise of spatial science and the emphasis on positivism in geography further polarized these tensions. Many in the spatial-science tradition paired the abstract with the supralocal and the concrete with the local (Agnew 1989, 129). Lukermann believed this polarization to be a characterization, not an analytic conclusion (Lukermann 1989, 61). After all, the distinction between concrete and abstract refers not to spatial scale but rather to the level of analysis (Duncan and Savage 1989, 195; Massey 1991, 271; Sayer 1991). As Massey points out, "thick description" of local areas involves complex issues of interpretation and requires a level of understanding that can be a theoretical challenge (Massey 1993, 147).

The human-land and regional-geography traditions have particularly been uncomfortable with the constraints of positivism that tend to reinforce these polarizations and have instead pointed to elements of uniqueness and understanding that go beyond those explored by a narrowly defined scientific method. This discomfort with the spatial-science tradition has recently come to the forefront of disciplinary debates with many geographers pointing out the problematic assumption of decentered persons living in geographic space (Adams 1995; Archer 1993; Daniels 1992; Massey 1993; Sack 1993). Contrarily, geographers such as Neil Smith lament this resurgence of what he identifies as neo-Kantian themes that privilege the individual and local over global or societal "empowerment" (Smith 1989, 116).

The work of two earlier prominent scholars in the society-land and regional traditions of geography, Carl Sauer and Richard Hartshorne, illustrates a longstanding unease with positivism within the discipline. Both found the dichotomy of the ideographic and nomothetic too simplistic a categorization and were uncomfortable with the positivist trends in the discipline that tended to reinforce this dichotomy (Lukermann 1989, 60).

Carl Sauer expressed concern over the methodological limitations of positivism. He wanted to counter the reductionist tendencies of social science that led to theorizing at the universal level (Williams 1987, 220) and feared that as rationalism and positivism dominated the work of geographers, the complex reality of areal associations was being sacrificed to a geography of causal relationships and deterministic formulas. He claimed that the interrelationships of objects, which together formed the landscape, constituted a reality as a whole that was not understood by a consideration of the constituent parts separately (Sauer 1963, 320-21). He saw quantitative geographers as misguided because they imposed ruling theories while ignoring the inherent diversity of nature and culture, the center of geographic inquiry (Entrikin 1984, 407).

Sauer looked for relationships among landscape "wholes" with each landscape having a recognized individuality. He claimed that this approach was still scientifically rigorous though not tied to simple causal relations (Sauer 1963, 322-23). Sauer used the method of multiple working hypotheses and temporary intellectual constructions to further his goal of discerning scientific facts (Entrikin 1984, 390). He avoided an explicitly causal vocabulary, but he continued to seek causes in the form of sequences of events rather than universal laws (Entrikin 1984, 406).

Price and Lewis argue that Sauer's historicism was antithetical to positivism, and that his work even anticipated the present theoretically open stance of postmodernism. His emphasis was on theoretical and methodological pluralism grounded in the inherent diversity of both nature and culture (Price and Lewis 1993, 11).

Like Sauer, Richard Hartshorne showed his open acceptance of diverse approaches to geographic inquiry and used terms, such as places and individuality, alongside methodology and theory and laws (Campbell 1994, 415). In his classic work, *The Nature of Geography,* Hartshorne described the tension between the nomothetic and the ideographic (Hartshorne 1939). He thought that the scientific enterprise of empirical observation, objectivity, concern for universality, and systematic ordering could not be reduced to the criteria provided by positivist philosophers of science (Agnew 1989, 123). He stated that even as the scientific ideal of certainty commanded that terms and concepts of description and relationship be made both as precise and as certain as possible—including striving for universals—that still did not exhaust the study of reality. He argued that there was always an individual remainder that was not described or explained and, if ignored, would lead to less than complete knowledge (Hartshorne 1939, 376-78). While he described geography as very much concerned with the study of individual phenomena (ideographic rather than nomothetic), he thought that a geography that was content with only the individual characteristics did not utilize the opportunities to develop universal principles. This was a view similar to that of Sauer (Hartshorne 1939, 382-83).

Whether true or not, Hartshorne was depicted by his opponents as the arch-typical idiographer. He was uncomfortable with distribution-and-process (cause-and-effect) studies in geography and thus ran into conflict with the growing positivist direction of the field (Lukermann 1989, 56). In reality, Hartshorne thought that causal explanations need not be restricted solely to relationships between general classes of objects or events as prescribed in logical empiricist accounts. He saw causal explanation as involving general mechanisms that produced observable outcomes, including individual or unique combinations of phenomena. Theory was thus no longer only associated with predicting a repeated series of events everywhere. Rather areal variation could provide a framework for examining the relationship between causes and outcomes without a presumption of universality (Agnew 1989, 125-26).

Hartshorne thought that positivist science could not capture the rich qualities of place. Geography needed something else. He saw associations in the landscape and the most interesting were those that differed from place to place. These still resulted from causal laws but resulted in unique associations that generalizations could not encompass without extreme oversimplification; these gave rise to specific or unique places that differed from the generic kinds of places that were the perceived outcomes of universal causal forces (Sack 1989, 155).

Place-Embedded Relationships

The discipline of geography and the society-land and regional traditions, in particular, have struggled with the importance of place as a creational feature of uniqueness and a relational element of reality. This appreciation of place and place-embedded relationships is being rediscovered (from geography's point of view) by many disciplines at present.

Scholars, spanning fields from economics to ethics, have rediscovered place and the importance of embedded relationships. At the same time, they call for a shift in approach to understanding society, community, and our relationships with the natural world. The field of geography and the Reformed Christian tradition—one that emphasizes the covenantal nature of the created order—have much insight to offer the present debate about the nature of place.

In present-day scholarship, the assumptions of the Lockean, Enlightenment worldview, are being questioned. This perspective still underlies much positivist social science (Skillen 1994). An alternative is presented to freedom as equivalent to the right to pursue one's self-interests: Freedom comes from a shared rootedness that in turn leads to fulfillment and full personhood. Not only do individuals gain true understanding of ecological systems, for example, agriculture in real places but also we become mature individuals only through the development of attachments to family, neighborhood, church, and other social institutions. Because these institutions find their expressions in a locale, there is a growing recognition that place is a contributing factor to personhood and that human freedom is only meaningful and fulfilling in community (Campbell 1990). It is there that our values are formed and grounded.

Geographer Nicholas Blomley has illustrated the tension within the Canadian legal systems between the Enlightenment worldview of individual freedom and an individual's commitment to a place (Blomley 1992, 238-39). He compared two court cases in Canada. In the first, the town of Kimberley, British Columbia, faces the closure of its silver, lead, and zinc mine. The residents are left with no choice but to move in order to survive, yet are reluctant to do so. They have a strong attachment to place, linked with a sense of shared history. In the second case, a doctor's right to practice in British Columbia is threatened by government regulation of physicians' licenses based on provincial need. One case concerns the right to move to a place, and the other with the right to remain in a place (Blomley 1992, 243-44).

The geography expressed by the court's decision is one that uses the language of space in a way more reflective of positivist geography as part of the spatial-organization tradition. From the legal perspective, geography appears as a plane of opportunity, a frictionless surface on which self-interested individuals make rational moves. A geography that recognizes ties to real places of uniqueness and relationship is missing (Blomley 1992, 246). But Blomley, like others who question this view of reality, asks: Is it possible to speak of individuals without attention to community and place? These scholars argue that personhood cannot be conceived of in the abstract individualistic terms of the courts and positivism. Furthermore, this abstract concept of geography as space does not mesh with the actual practices of the doctors who sought the freedom of mobility. They did not want to move in space, but wanted to be able to choose *particular* places (Blomley 1992, 247-48).

Similar criticisms of these positivistic, Lockean assumptions underlying orthodox economics, and thus economic geography within the spatial organization tradition, have been forthcoming. Like the spatial account of human existence that underlies the legal system, neoclassical economists perceive individual choice and pursuit of self-interest as basic to society and individuals as its building blocks (Clemenger 1994). Such persons are necessarily "disembodied, dispassionate, defamilied, and degendered" so as to fit this unencumbered model of humanity (O'Neill 1994, 41).

Granovetter argues that this atomized and unencumbered view of human action fails to include the concept of trust among individuals. Trust implies relational embeddedness that is chosen and developed over time, leading back to an alternative concept of freedom arising from longstanding relationships (Granovetter 1985, 489). The kind of trust and embeddedness that Granovetter describes can only be based on proximity, on place and community, and recognizes the reality of freedom arising from relationships.

Bennett Harrison attempts to illustrate the importance of community proximity and relationships in his work on prosperous new industrial districts. He claims that their existence is not explicable by conventional neoclassical economics (Harrison 1992, 471). Harrison explains the growth of these areas by the emergence of informal ties based on trust that arises out of experience. This experience is built up over time, through continual contracting and recontracting, informal deal making, and assistance to one another in times of stress. In the standard space-based urban-economic models used in spatial analysis, proximity facilitates the exchange of information on which individual atomistic decision makers may then act. The place-based argument offered by these industrial district theorists is built on a different logic: Proximity digests experience, which leads to trust, promotes recontracting, and ultimately enhances regional growth (Harrison 1992, 477).

Positivism has led scientists to seek explanations that crossed localities, interested not in complete understanding of a specifically situated phenomenon but in partial understanding of widely dispersed but similar phenomena. Such scholars investigate the production of mobile information that is applicable through any change in spatial or social location. In contrast to "scientific" knowledge, the new emphasis on "local knowledge" asserts that understanding may be inseparable from a particular place because it is embedded in the natural features and labor processes of that place [environmental and social embeddedness] (Kloppenburg 1991, 529, 537).

Feminist theorists have attempted to articulate what such a place-based, relational science would look like. Such objectivity emphasizes situated knowledge, limited location, and the inability to split subject and object (Haraway 1988, 581). Furthermore, Evelyn Fox Keller claims that a positivist approach to science does not allow questions about human relationships to natural systems. She describes an alternative that has the goal of understanding rather than prediction, and reflects as well as affirms our connection to the natural world (Keller 1985, 147, 166).

All of these arguments sound much like Granovetter's explanation of economic embeddedness, of local knowledge, and scholars' arguments for freedom arising out of relationships to place. These criticisms of the emphasis on universal space are similar to the claims of Hartshorne. He said that both history and geography might be described as examining reality from a naive point of view, looking at things as they are actually arranged and related; however, "hard" science, the alternative, required phenomena to be taken out of their real settings, at the cost of context (Hartshorne 1939, 373). But does this contextualization of knowledge and of human society lead to a lack of any common universal experience, understandings, or knowledge? Before this question can be answered, some discussion about the role of theoretical presuppositions is needed.

Presuppositions and Worldviews

What are these critiques of the accepted assumptions about the nature of society, humans, and the natural world telling us about the reality of our existence, and the creation order within which we live? Ultimately one must recognize that, contrary to the positivist perspective, rational thought does not overcome differences in assumptions about the nature of humankind. In fact such a belief in the neutrality of rational thought is itself a religious commitment (Newbigin 1989; Clouser 1991).

The most basic commitments are at the heart of any theoretical perspective, yet the spirit of positivism would have us believe that truly rational and scientific thinkers assumed nothing except one or two self-evident propositions. Assumptions are beliefs arising from ultimate commitments (Hart 1984, 328).

No coherent thought and scholarship are possible without taking some things as given. These ultimate commitments, from which worldviews are developed, can only arise within the context of a community or a tradition. Peter Berger calls these societal assumptions "plausibility structures." These traditions determine which beliefs within a given society are plausible to its members and which are not. With a dominant positivist plausibility structure, Western society looks to reason as though it were an independent source of information alongside tradition or revelation (Newbigin 1989, 8).

"Communities of commitment," as Bellah calls them, take on their own perspectives (Bellah et al. 1985). These communities cannot be understood through the method of science, which tends to reduce values to the level of individual preference and then attempts to predict them on the basis of measurable socioeconomic variables (Miller 1993, 48). David Hummon (1990, 11) characterizes these belief systems or perspectives as unconscious and community wide. He says that though they are expressed by individuals, they are public in that they are learned and sustained in the context of relationships with others in a community. If that is true, then it follows that the cultural order is not reducible to the individual phenomenon, something pointed out by both Sauer and Hartshorne.

Philosopher Roy Clouser describes the inherent problems of positivism that insist on this reductionism of reality. It asserts that some aspects of reality are basic to all the rest. These most basic aspects have a decisive relation to all the rest, a connection that, from a Christian perspective, is the same as the relationship God has to creation. As a result, these aspects are viewed as divine. Clouser argues that this is a pagan-like idea of god, "for apart from pagan religious belief, what would be the reason for trying to construct a theory of reality by reducing all the rest of its aspects to some one or two?" (Clouser 1991, 173).

In this respect at least, radical geography is the same as positivist geography. They are both reductionistic. The only difference is that Marxist geographers complain that positivist geographers either lack theory or fail to grasp the factors that are truly the most basic—factors of class structure and political economy (Peet 1977). For example, the "restructuring" approach in radical economic geography has treated place as the result of economic processes and relationships such as spatial divisions of labor (Day and Murdoch 1993, 88).

A reductionistic worldview limits a scholar's perspective. Part of the reason for this is the intertwined nature of reality. All aspects of reality are found in every state of affairs. Thus, all things can be viewed and, in some sense, explained from the standpoint of virtually any of them. In this way, reductionism becomes self-confirming and tends to blind its adherents to the fact that the same sort of confirmation is possible from the point of view of other aspects of reality. This is why the issue of ultimate commitments is crucial. Thus, the belief that any aspect can exist independently is just as much a limiting idea as the belief that it is God alone who is divine (Clouser 1991, 191, 195).

A Reformed Biblical Perspective

The Bible's presuppositions and its doctrine of creation lead in a different direction than positivism. If one confesses that God has created all, then it follows that everything is dependent on God alone for its existence; thus denying that status to anything else (Clouser 1991, 173). Hence, no matter how hard scientists try to isolate aspects of reality, all of them continue to display unbreakable connections to all the others. This reality of interconnectedness is evident in the fact that while positivist science claims that any aspect of reality is capable of existing independently of the others, Christian scholars cannot conceive of reality as so constituted (Clouser 1991, 173, 193).

Positivist science associates rational knowledge with certainty, truth, and justification. This rationalism in turn is founded on reductionism and the independence of individual aspects of reality. In contrast, the biblical tradition is that knowing is usually a function of relationships (Hart 1984, 356), an emphasis that was likewise evident in the place-embeddedness scholarship described earlier. After asserting the truth of God, Scripture is mainly concerned with clarifying the ways in which God relates to us. God has made this the model of what it is to be rightly human, which means being in relationship (Clouser 1983, 394, 405; Brinsmead 1978, 8). Humans are related to God, to the community, and to the created order.

In support of this emphasis on relationship, Douglas Hall argues that one of the basic or foundational biblical understandings is the concept of humanity as the image of God. While traditional theological reflection has centered on traits possessed by humans that image God, he suggests that a minority tradition has identified the image of God not as a quality of being but of relationship (Hall 1988, 12-13).

Central to this recognition of commitment, community, the whole, and the relational aspects of reality are the biblical theme of the covenant. Robert Bellah has pointed out the conflict between the Lockean, reductionistic notion of society and the Hebrew notion of covenant in the biblical religion. The covenant is a relationship between God and a people, but the parties to the covenant, unlike the parties in the Lockean contract, have a prior relationship: the relationship between creator and created. And the covenant is not a limited relationship based on self-interest, but an unlimited commitment based on relationships of loyalty and trust. It involves obligations to God and the neighbor that transcend self-interest though it promises a deeper sense of self-fulfillment through participation in a divinely instituted order that leads to life instead of death (Bellah 1990, 11). In recognition of this human quality, economic anthropologist Karl Polanyi argued that interdependence was as intrinsic to society as was the individuality of each person (Woods 1994, 248).

John F. A. Taylor has also seen how the reality of covenantal relationships runs counter to our society's positivist perspective. Though he sees such relationships as indispensable to any understanding of societal structures, the inability to analyze them by the scientific rigors of proof cause them to be viewed by positivist science as arbitrary conventions, fictions made up by self-interest. Yet, they are fundamental to our understanding of ourselves, our worldviews, and our society. Taylor draws on biblical images such as Job 29:14: "I put on justice, and it clothed me." In that phrase Taylor sees the whole burden of the Hebrew's sense of history: In community, he is clothed; cut off, he is naked; and there is no other nakedness (Taylor 1966, 7, 23). Jewish theologian-philosopher Martin Buber argued that only in primal relation do people become persons, and that we need to recognize the profound reality of interrelationships expressed in a sense of covenant between persons (Van der Hoeven 1990, 26, 29). Granovetter and Harrison thought such a sense of community could only be realized in a place, where day-to-day relationships could develop.

The reality of covenant offers an alternative to the focus on the individual and the view of collective life as fictions of rationalized interests. Yet, liberal theorists claim that such a covenant order is a primary or primitive institution because it is familiarized and gendered. This ideological hierarchy of contract over covenant is assumed in all the social sciences (O'Neill 1994, 39, 51).

The spatial organization tradition developed on this same assumption of the de-gendered, rational individual. Fred Schaefer reflected this orientation when he argued that "the connections between ideology and political behavior, or the lawful connections between the psychological traits of a population and its economic institutions do not concern the geographer" (Schaefer 1953, 228). And Walter Isard, father of regional science, differentiated between "pure" theory that could be used in social planning and "realistic" theory that considered institutional forces (Spate 1960, 385). Spate (385-86) pointed out, "But on what definition of 'social' can one legitimately leave out institutional forces, since society is just that—institutional forces?" The danger is that the abstraction is regarded as more "real" than reality.

Are the images that arise from the idea of covenantal community somehow reflected in what Sauer and Hartshorne and the more contemporary critics of positivism were trying to describe? Kuyvenhoven, for example, writes on the movement back and forth between the personal and the communal as found in Psalms 25 and 26. He points out that personal sin affects a whole community and that when a group becomes an incidental gathering of individuals, it lacks the power of a covenantal group. He goes on to argue that it is God's desire to reestablish community among people in order to make them whole beings. Only within this community does the individual rightly develop (Kuyvenhoven 1974, 46-47). This sounds very much like some of the critics of positivism.

How does this vision of covenantal relations connect with geography's emphasis on society-land connections? As Hall recognized, and as Keller and Sauer also emphasized, the relational quality of the created order includes the natural world, an insight geographers have long appreciated. Hall argues that this relational characteristic of our being describes our unique calling: to be in responsible relationship with God, each other, and the rest of creation (Wilkinson 1991, 285). I believe that the biblical image of covenant that focuses on the covenanters *with their households*, gives us some clues (Kline 1972, 78).

In the Bible, a covenantal document has as its function the structuring of the covenanted kingdom. In this connection, the imagery of God's house comes to the fore and represents victory over chaos. In Exodus, this house-building is of two kinds. First, the covenantal words of God spoken at Sinai structure the people Israel themselves into the formally organized house of Israel. Second, a more literal building for God's habitation—the tabernacle—is constructed (Exodus 40:34-38). In these and other instances, wisdom is associated with house-building (Proverbs 8:22–30) and this implies the presence of a covenant community (Kline 1972, 79-80, 86, 90). House-building, wisdom, and community converge.

The Christian faith, while it has a strong element of personal commitment, demands that we soon move to the level of community, taking care of each other. This community, especially expressed in the image of the body of Christ, is not a building contracted by individuals acting out of self-interest; it takes on a life of its own. Suggestions of what this means are found in the New Testament, such as the well-known word *oikonomia*. In its biblical context, it implies responsibility, care, and acting on another's behalf. A related term, *oikonomos*, means steward, or the steward of a household (Goudzwaard 1992, 5). Again we find these associated concepts: house(hold), managing wisely for the good of the community (household). This is clearer in Luke 12:35-48. The wise steward is the one taking care of the household while he waits for the master. He doesn't settle into complacency, eating and drinking, thinking only of himself. Rather the wise steward gives each member of the household his or her portion of food at the proper time. He is busy taking care of the household.

While the New Testament implies a certain stewardly attitude toward resources and the rest of creation, the Old Testament in particular ties body and household images, or covenantal relationships, concretely to land resources. The Israelites were given land by God to be kept as long as they were in a covenant relationship with Him. When they repeatedly mistreated the land and/or treated their fellow community members unjustly, they were finally banished from community and the land. Thus, individual ownership was affirmed but the right to the land was directly tied to one's relationship to God, the fruits of which were seen in the concern for brother or sister and for the land itself. The law of jubilee (Lev. 25:10) guarded against speculation or the removal of land from these relationships and concerns. Again the connections are there: covenant, household, and housebuilding, in association with wisdom and stewardship (taking care of the whole and the land). Humans and land are part of an intertwined whole whose expression is found at a spatial level where covenantal relationships can form.

Conclusions

Positivist assumptions about human nature and knowledge have failed to accurately describe reality and thus, positivist foundationalism has been found wanting. This acknowledgment of positivist science's failures has created an opening for a different direction (Goudzwaard and De Lange 1995, 93). Reformational, or neo-Calvinist, philosophy has been no less critical of the fundamental notions of modernism than postmodernism. As illustrated in this chapter, themes of embeddedness in nature, and the relational aspect of personhood are similar. However, Christian perspectives do not lead back to an absolute relativism as do postmodern perspectives. A biblical faith sees the source of diversity, relationship, and nature, as the God who has provided a stable and reliable order for his creation (Botha 1995, 163). The neo-Calvinist Christian tradition in particular, has placed the notion of the creational order as a central concept for understanding the diversity of social structures and their interrelationships, recognizing the relatedness of all forms of diversity to this transcending point of reference (Botha 1995, 168, 171). This acknowledgment of diversity stands in opposition to the search for an objective and universal science as pursued by Enlightenment thought (Griffioen and Verhoogt 1990, 10-11).

An emphasis on place or context does not eliminate universal normativity. Griffioen and Verhoogt go so far as to say that contextualization must answer to norms or it cannot serve as a countervailing power to the rationalism of Enlightenment thought (Griffioen and Verhoogt 1990, 13-14). Thus, while a biblical perspective acknowledges both the unique and the universal, the ideographic and the nomothetic, it also acknowledges a source of this diversity. What humans experience as order, pattern, structure, coherence, and regularity in the world is a concrete expression of God's will for an ordered and coherent world. And what humans experience as individuality or the unique is an expression of the rich diversity of the creatures, each with its own unique calling, all known by God and by us as irreducibly real in their own right (Hart 1984, 204). In the case of place, the universal themes of covenant and commitment are crucial for our understanding the nature of the local expression of this reality.

Chapter 5

God's Own Countries? Contours of a Christian Worldview in Geography

Gerda Hoekveld-Meijer

Prolegomena

HAVING BEEN ASKED to incorporate a Christian worldview into twentieth-century geography education, I ask myself whether there are Christian concepts that can be incorporated into geographic theory, and, if there are, whether their influence would result in a Christian geography that can be distinguished from its non-Christian sibling. For me, a geographer and theologian, this Kuyperian question is intriguing (see below, Aay, chap. 7, Griffioen, chap. 8). It is the more intriguing because the Free University, established by Abraham Kuyper, did not consider geography a discipline worth keeping after its already very late establishment in 1961. In 1985, the Department of Social Geography and Planning, a Ph.D. granting department, was transferred to the secular University of Amsterdam. However, a small geography presence remained in order to serve the requirements of teachers. Educational geography became a small section within the Institute of Didactics and Education (IDO). But even this small cell was too heavy a burden to maintain for the university. So today, the Free University, an educational institution with more than 14,000 students, has no human geography faculty within its walls.

In considering the demise of geography at the only Protestant university in the Netherlands, my interest in the relationship between Christian worldview and geography is colored, of course, by the question of why the board of a Christian university did not value social geography and planning. Is it because the dimension of place and space—so important in geography—has no self-evident relevance in a Christian worldview? Or is it because since the days of Adam Smith, Marx, and Weber, the dimension of time came to dominate the dimension of space in social theories? Is that why the Christian philosophy of history of Herman Dooyeweerd, according to one sympatic critic, is focussed on the progressive character of social change or cultural development that unfolds itself in a unilinear temporal order of stages (McIntire 1985, 103–117)? Are we to accept Kuyper's idea of cyclic notions of regeneration (*palingenesis*) and being begotten anew (*anagenesis*) (cf. Wolterstorff 1989, 59) as progressive in the sense that each cycle is better than the preceding one? Do we really believe the postmodern worldview in which "space compression [is] generated out of pressures of capital accumulation with its perpetual search to annihilate space through time and reduce turnover time" (Harvey 1990, 306-7) and, that in response to this time-space compression, parochialism, myopia, self-referentiality, and an alarming irresponsibility create new borders on the local level (350-51)? If we do not accept these trends, we have to ask ourselves whether in a Christian worldview the concept of space belongs to that "mix of imagination and conceptuality" that "functions as a map" and sets out "a path for meaningful action" (see below, Griffioen, chap. 8).

In view of the insignificance of space in postmodern worldviews, it is not enough to apply biblical concepts to geographical theory. In my opinion, we have to reconsider the concept of linear time and the undisputed dominance of the concept of *development*, the goddess of social science. So I suggest we remove this goddess from her pedestal and replace her with the theologically laden concept of *regeneration*. I also propose to consider development from the perspective of circular and stationary time instead of the familiar perspective of linear time in order to see how the concept of space fits into this Christian "mix of imagination."

Regeneration versus Development

In his contribution to a conference organized by the Free University, Calvin College, and the Institute for Christian Studies (Toronto), Van der Hoeven underscored the duality of the concept of development in the light of *encounter*. With respect to natural development, the emphasis is on its status as a "given," hence a certain predeliction for circular models. Nevertheless, cultural development tends to be associated with "man-made," technical, and linear processes. The very concept of development is marked by this duality (Van der Hoeven 1990, 25). However, Van der Hoeven stresses that the same duality also "makes us aware that our existence is full of purpose rather than a merely linear or circular process." But many man-made developments follow the curve of natural growth with a distinct beginning and definitive end. The circular shape of development, whether it is a natural or man-made process, is only conceivable as such if we acknowledge that development is place bound (Gurvitch 1963, 177). When we ignore place, development is a curve with an asymptotic tendency in one of the two directions or destinies indicated in Figure 5–1a. It may dip into oblivion (line A) or the exponential growth turns into a process of involution (line B) because space is limited in contrast to time and development. When we consider development from where it takes place, the asymptotic curve turns into a circle with or without spilling over into the neighboring space (Figure 5–1b). Although the curves in Figures 5–1a and 5–1b look different, they show the same process from two points of view.

Figure 5–1a
Time perspectives in relation to development and place:
Natural development (A) and cultural development (B)

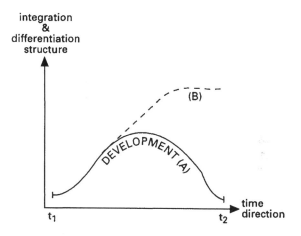

Figure 5–1b
Time perspectives in relation to development and place:
Development from the perspective of place

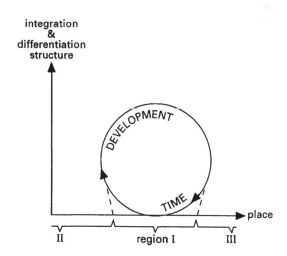

In contrast to Van der Hoeven who drew attention to the duality of time, Dooyeweerd viewed cultural development as a strictly linear historic process.

Summarizing Fernhout's discussion on the topic (Fernhout 1990, 88–94), Dooyeweerd assumed that this process was characterized by a direction and structure. The structure is defined in terms of an ongoing differentiation and integration, while time is predominantly seen as direction. The arrow of time points in a definite direction determined by the dialectics of Western culture (but see McIntire 1985, 103–117, and Griffioen 1986, 84 who are rather critical of this view). But the flow of time is conditioned by the limits of space (the earth). Because culturally there is no end to technical differentiation, the historic unfolding of ontic structures postulated by Dooyeweerd resembles the curve of involution (line B in Figure 5–1a). If, however, Van der Hoeven is right in assuming that the duality of time has a purpose, we had better introduce a variant of time that includes the two perspectives and gives us room to maneuver between the fatalistic perspective of natural development, and the optimistic view of involution that is inherent in the Western belief in the technical side of cultural development. As far as I can see, the biblical concept of the kingdom of God offers us a place to maneuver because it is characterized not by development but by justice (cf. Marshall 1986).

According to Wolterstorff (1990, 160–163) justice and rights (morally legitimated claims), are the same phenomenon (contra Marshall 1990, 153), who claims that rights are the consequence of justice). But, in the kingdom of God, the concept of rights has no meaning. Whatever one has is given. If Wolterstorff is right in equating justice and rights, the main difference between secular kingdoms and the kingdom of God is that in the latter justice exists without rights. However, justice cannot exist without duties. There are fundamental human duties that can be derived from the first and second commandments. As example, I point to the duties that imply a situation of injustice and prepare the way for regeneration: the duty to remove idols (including the goddess of development) just as Gideon did (Judges 6:25), and the duty to blot out the memory of Amalek, the prototype of people who steal from the poor and kill their brothers (Deuteronomy 25:17-19). So, in this chapter, I shall use the term justice in the context of the kingdom of God as a condition in which people assume their duties instead of asserting their rights.

Kingdom of God is not a metaphysical concept but a metaphor of a society ruled by righteous leaders. Given our interest in justice, I would like to draw attention to Houtman's *World and Anti-world* (1982). In his discussion about the relationship between people and environment in the Bible, Houtman (1982, 82–93) demonstrates that the kingdom of God cannot be conceived in accordance with a mechanistic worldview or, rather, a mechanistic world picture, following Griffioen's distinction between these two concepts (Griffioen, below, chap. 8). The kingdom of God is based on the basic belief that YHWH is king and that nature is his servant. The sun and the moon create conditions to support the natural life cycles on earth (Houtman 1982, 80); heaven and earth are summoned to be witnesses in the lawsuit of God against people who do not pursue justice (76). With God as judge, nature is his instrument or "hand of God" (50–59). Justice is not a variable that progresses in time. It is dichotomous in character, like regeneration. There is justice or injustice; regeneration or degeneration. These contrasting modes of society are illustrated in the curve in Figure 5–1c.

Figure 5–1c

**Time perspectives in relation to development and place:
Development viewed as a series of historical events from the perspective
of justice**

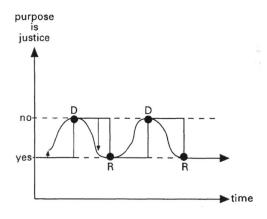

A just society degenerates into an unjust society the moment (t_1) people forget their fundamental human duties, although they get a chance to regenerate. This happens when they remember their duties and act correspondingly (t_2). The curve in Figure 5–1c is punctilinear, meaning that time is created by events. The sequence of degeneration and regeneration suggests that the kingdom of God does not necessarily represent the final stage in every kind of development (cf. Perrin 1963, 159ff, 184ff.). It *may be* the last stage depending on whether all future generations can maintain justice. If not, these generations will relapse into a degenerative stage. However, they can be regenerated. When we realize this, the crenel-like curve of stationary development depicted in Figure 5–1c is, in fact, a series of time cycles fluctuating between the two dichotomous levels of justice and injustice. Contrary to the belief of Dooyeweerd that cultural development is a historic progress that opens up structural possibilities, it is more like a ball hitting the ceiling of injustice created, among other things, by a proliferation of rights, and then bouncing back to the street where people encounter each other and are fully aware of their duties.

The recurring collapses of Israel in the biblical narratives supports my thesis that there is no development in our Western sense, nor a final stage, only degenerations followed by exiles, and regeneration (Hoekveld-Meijer 1996, 15-25). As brother, Israel is exiled to Aram (Gen. 25-28). As father, Israel is exiled to Egypt (Gen. 37-50). As nation, the sons of Israel are exiled to Egypt and Aram (1 Sam. 2 Kings; Jeremiah 40-45). There is no progress in the history of Israel, there is only a repetition of collapses. But time and again the people were invited by prophets and narratives to prevent exile, which is the ultimate form of political degeneration. They only had to return to the law and pursue "justice and justice alone" (Deuteronomy 16:20).

Now, regeneration is one of the concepts that Kuyper rightly saw as the essence of Christian religion because of its close association to the concepts of redemption and deliverance. There is no regeneration if there is no redemption. Kuyper associated regeneration with people and concluded that there are two types of sciences because there are two kinds of scientists, regenerated and nonregenerated ones. Although this a priori conviction is contested by Wolterstorff (1989, 58-66), I am inclined to agree with Kuyper that, at least in social science, the two groups of scientists have different premises. For secular thought the concept of regeneration implies a fusion between the concepts of development (rise) and decline (fall). A decline is necessary to maintain conditions in which the continuation of matter in situ is secured. Christian scholarship, though, starts with another premise. Christians believe that a decline can be prevented, whether one sees the coming of the kingdom of God as final or as a recurrent event. The fall can be prevented because regeneration is possible in a redemptive encounter. By this, I mean an encounter in which unrighteous leaders ask their victims to redeem them and victims indeed redeem their oppressors. In this encounter between two socially unequal parties the unrighteous party is regenerated, while the victims are delivered. If, however, the unrighteous do not ask for redemption, their victims will be delivered by people who carry out their fundamental human duty. So instead of associating cyclical and linear development with fate and a fatalistic belief in the rise and inevitable fall of empires as is usually the case, the concepts of redemptive encounter, deliverance, and regeneration reveal that these two curves of development can be converted into a stationary development (a steady state).

When I consider stationary development as the destiny of a righteous society, this does not mean that the past and the future have lost their meaning.

On the contrary, stationary development is a dynamic equilibrium requiring a permanent dialogue with the past (cf. Van der Hoeven 1990, 32) and the future (cf. Zuidervaart 1990, 39) as an extrapolation of the past and with the present, which is embodied in generations of neighbors, each within their own land. It is not the destiny of empires to decline, once they have appeared. It is the freedom of their leaders to decide which of the four curves of development should be their guide. In this sense Layendekker (1986, 159, 164) suggested that we should be sensitive to the variations of time perspectives, and that we should redefine development in terms of improvement. As can be seen in Figure 5–1c, improvement has two modes. One consists of measures that prevent a development from reaching the level of injustice. The other improvement takes the shape of measures that are meant to restore justice. It is clear that the longer a society pursues injustice by insisting on rights, the more draconian the measures that must be taken in order to restore justice and prepare the way for regeneration.

The next step in the dethronement of the goddess of development and linear time, her faithful servant, is to explore how the concept of space fits into the conceptual interrelationships among regeneration, justice, encounter, and the perspectives of stationary and circular time. The concepts of externality field and regional inequality reveal the spatial dimension of a Christian worldview. I shall first establish the connection between regeneration and externality field.

Regeneration and Externality Field

Figure 5–1d depicts externality fields or areas where neighbors experience the negative effects of their development on each other's territory (see also below, G. A. Hoekveld, chap. 6). In general, the roots of these negative effects are found in the past (the father in Hebrew imagery). Justice is achieved when the sons (the present) take measures to minimize these effects so that their sons (the future) may not experience a political collapse. Hence, the concept of externality field is a place of encounter with a historical dimension. It is the place where three generations meet each other. Negative effects are driven by an ongoing differentiation and integration that produces unjust cultural development. Hence, externality fields are the products of cultural development and social injustice. Gerben de Jong—the only Dutch geographer in a university setting who raised the question of how the Bible and human geography are interrelated—would have called these fields examples of geographical areas (regions) because they are characterized by an areal totality of integrated differentiations of the phenomena of nature and culture (de Jong 1958, 100). In his view, "this differentiation arises from the human-land relationship and is the object of geography" (99). But, as I said above, externality fields arise from the human-to-human relationships that, of course, imply human-land relationships. By insisting on the traditional, geographical paradigm of society-land relations in which borders are observed as the outer limits, De Jong missed the moral dimension of borders as divides associated with externality fields. Without that dimension regeneration has no meaning. But what is the connection between these two?

Figure 5–1d
Time perspectives in relation to development and place:
Externality fields of regional developments

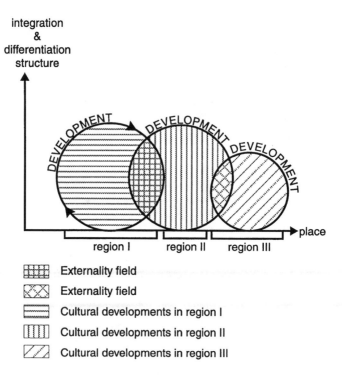

▦	Externality field
▧	Externality field
▭	Cultural developments in region I
▥	Cultural developments in region II
▨	Cultural developments in region III

Integrated differentiation stands for spatially overlapping modalities or spheres such as the overlapping circles of development in Figure 5–1d. De Jong overlooked or ignored some implications of the political sphere that the other modalities do not have. In other spheres, property is mobile (it can be carried, moved, lost, sold, put away) but in the political sphere property is immobile: Land nor the place-bound negative effects can be moved. The immobility of the political sphere and the place boundedness of the negative effects emanating from other spheres have two implications. The first is that autonomy only has a limited meaning because spillover effects are rooted in other spheres and in the territory of other people. The second is that negative externalities are a spatial form of stealing. Negative externalities spoil or take what is good from the land of neighbors. It is a fundamental human duty of the people who are responsible for the negative externalities to return what is stolen, restore what is ruined, and, above all, remove the origin of the negative effects. The last is the most difficult because it requires a regeneration of the culture in the country of origin. In short, the connection between regeneration and externality field is the concept of justice; it is realized when people do their fundamental human duty and take care of their neighbor's backyard by improving their own, for in the kingdom of God there are no externality fields. There is, however, regional inequality. So the next question is how to relate regional inequality with justice and regeneration.

Regeneration and Regional Inequality

Justice arises when people do their fundamental human duties such as the removal of unsustainable economic growth and suppressive political systems. Hence, justice requires strong brothers (or firstborns). They are a theological necessity to secure a condition of justice. The concept of the strong firstborn as well as its alter ego, the weak lastborn, implies political inequality that is opposite the "levelling universalism" and the "equalizing generality" that characterizes Western thinking (cf. Van der Hoeven 1990, 32-33 who states that "this biblical theme has been recognized insufficiently as a principal motif"). In the kingdom of God, the negative aspects of political dominance are neutralized because dominance is defined as service (as ordained by YHWH in Genesis 25 and by Jesus in the New Testament). Such service, which is a fundamental human duty, does not amount to a trickle-down effect as Zuidervaart (1990, 41) fears in a society in which the goddess of economic growth is not dethroned. It is the duty of a strong neighbor to deliver weak ones from negative externalities caused by economic growth.

Now, inequality is not a popular concept. It is contested, not only by Western philosophers in our age, but also by Ezekiel, the postexilic prophet-priest who was a socialist before his time and lived in Babylon. He despised Joshua's spatial organization of Israel after it returned from Egypt because of its striking regional inequality. As the biblical polemic bears relevance to our theme of how the concept of regeneration can be meaningfully related to regional inequality and justice, I compare Ezekiel's and Joshua's accounts.[1]

In the last chapter of his book, Ezekiel envisaged an egalitarian society where each tribe receives an equal share of land (Ezekiel 48). The land itself is ruled by one special family of Levites, the Zadokite priests who live in the "Sacred Reserve" located on "the highest mountain of Israel": Mount Ebal, not far from Nablus (near ancient Shechem opposite Mount Gerizim, see Map 5–1). Although they do not hold it as possession (Ezekiel 44: 28), they control Israel from this region. "He who controls Shechem, controls Canaan" is a common adage among archaeologists in view of the continuity of the road net-

Map 5-1
Division of postexilic Israel into thirteen zones according to Ezekiel

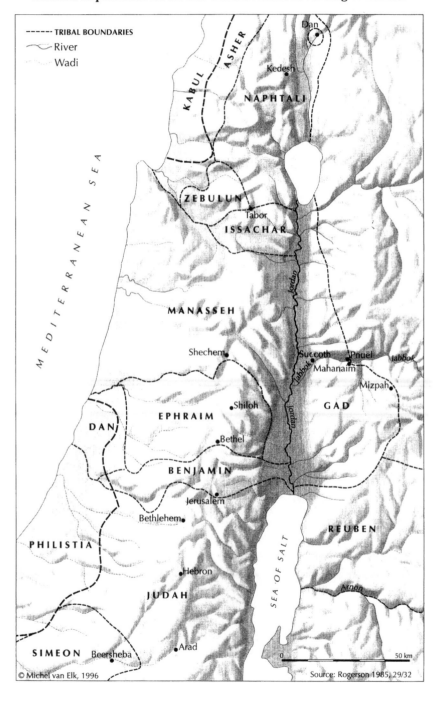

work and archaeological remains dating from the nineteenth century B.C.E. until Roman times (Dorsey 1991). The other Levites inherit the best of the land (Ezekiel 17: 22-23; Ezekiel 48). Because the tribe of Joseph is represented by his two sons and the Levites inherited land, Ezekiel divided the land of Israel into thirteen provinces. In his blueprint of the sacred reserve, Ezekiel designed a big temple city in the style of a Babylonian Baal temple, which he separated from a much smaller capital city by a five kilometres wide zone of Levite land. He pushed the king from the political center to the periphery (Ezekiel 45, 48: 8 -22, especially 48: 21) in order to prevent future kings from oppressing the people (Ezekiel 45:8). The function of the king was reduced to that of high priest (Ezekiel 45:17).

Ezekiel's blueprint can be read as an intended alternative to Solomon's division of Israel. Just as Ezekiel, Solomon divided Israel into thirteen provinces. Twelve provinces had to serve Judah, the thirteenth province. This division was based on the egalitarian principle of taxation and on colonial rule. Each month one of the twelve districts had to provide the king and his horses and the mayor of the main port, Tyre, with food (1 Kings 4:26-28; 5:11). So each colony had to pay tax in equal measure to the motherland, Judah. Given the postexilic dates of the prophecy, the map of Ezekiel with Shechem as the planned epicentre of the new Israel is a silent witness to an ancient controversy about the function of Jerusalem. Apparently, Ezekiel rejected Solomon's Jerusalem. But he also rejected the brothers who returned from Egypt (Ezekiel 20:1-26). He despised them for their laws that would not lead to life, and their religious practices (Ezekiel 20:25-26). So he must have despised Joshua's division as well, for that was based on these same despicable laws. And indeed, these maps are based on opposite principles: The map of Joshua is based on inequality, that of Ezekiel on equality.

Joshua divided the land into eleven unequal portions (Map 5–2). Another major deviation from Ezekiel's map is that Levi does not receive land. The Levites may live in cities and use their common grounds (Joshua 21). From the writer's lengthy description, it is nevertheless clear that each district (province, tribal land) is assigned by lot (Joshua 13–19). The tribe of Joseph, the brother who, on instigation by Judah, was sold to Egypt (Gen. 37:26), gets three portions all together, and is allowed to take more from the wooded hill country, the lands of Judah and Benjamin his younger brother by the same mother (Joshua 17:14-18). Joseph's son Ephraim receives Shechem, the metropolis. Manasseh gets land on either side of the Jordan north of the Shechem pass. Ephraim and Manasseh get a double share because they are blessed already (for they are numerous, Joshua 17:17-18). Levi, Simeon, and Judah, on the other hand, lose their share of land, or part of their allotted share because they ignored some fundamental human duties. Levi and Simeon broke a covenant with their brothers, the inhabitants of Shechem by killing them after they returned from their first exile (Gen. 34). Levi has to live in cities (Joshua 21). Simeon has to live within the tribe of Judah, his younger brother

(Joshua 19:1-9). Judah, the tribe of David that brought forth so many oppressive kings, as is testified by the prophets, sold Joseph his brother as slave instead of protecting him from his other brothers. So Judah has to share his land with three brothers: Levi (in his cities) Simeon (in the plain), and Joseph (in the hills). Instead of Ezekiel's plan of regeneration, which is based on equality and the rejection of the concept of historic guilt that ought to be restored by the innocent generation (Ezekiel 18:1-3), Joshua's plan is unmistakably based on the principle of inequality and the recognition that historic guilt must be restored to the extent that in some cases political roles are reversed. The unrighteous big brother is regenerated as the small (last) one. The rejected (small) brother (Joseph, the last) is reborn as the first (leader) in the person of his two sons who rule Israel from Shechem, its most contested crossroad. If social injustice is the condition of the kingdom of the big brother, this is paid for by loss of land and rule! This is more than the trickle-down effect of development aid. If a redemptive encounter of strong and weak brothers does not take place, regeneration means revolution. When the redemptive encounter takes place, the old map of regional inequality based on injustice is also transformed. The new map shows a regional inequality based on justice. In other words, regional inequality is the spatial dimension of regeneration if it is the consequence of a restoration or overcompensation of past injustice.

When we consider the concepts of regional inequality and externality field as spatial expressions of a complex of theological concepts such as regeneration, justice, and encounter, it follows that the concept of border must be meaningful as well because borders delineate all these regions and fields. In the next two sections, I shall demonstrate that borders are a theological necessity in connection with regeneration, and with places of encounter and justice. As such, borders are the spatial expression of these concepts.

Borders and Justice

Apart from the utopian character of a regeneration and the subsequent righteous regional inequality, Christian scholars may have problems envisaging how borders can be related to encounter and justice. The kingdom of God, after it has reached the ends of the earth, is thought to be a place where all borders have vanished. It can be argued, though, that even the kingdom of God cannot exist without borders. In the biblical narrative, each political collapse (degeneration) results in an exile. After that, the people of Israel are delivered and allowed to return to their point of departure. The completion of degeneration as well as of a regeneration end with a border crossing. In other words, borders are a theological necessity because there must be at least one other land to accommodate a degenerated generation during their exile. The next question is how borders relate to justice.

Map 5-2
Joshua's division of postexilic Israel in 480 B.C.E. into twelve districts according to Joshua 15-19 and the traditional geography of Canaan

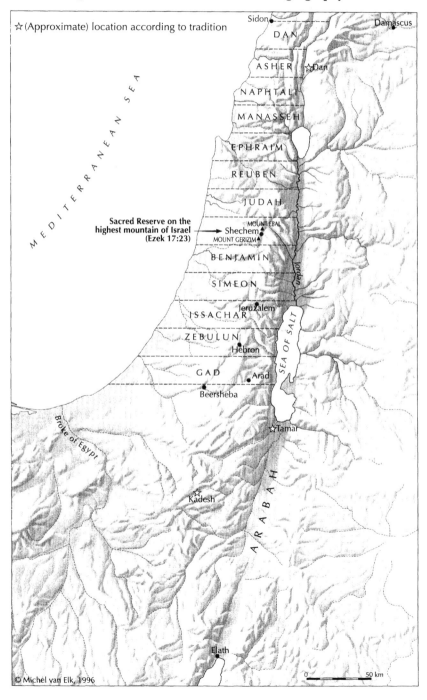

☆ (Approximate) location according to tradition

Sidon

Damascus

DAN

ASHER ☆ Dan

NAPHTALI

MANASSEH

EPHRAIM

REUBEN

JUDAH

Sacred Reserve on the
highest mountain of Israel → Shechem
(Ezek 17:23)

MOUNT EBAL
MOUNT GERIZIM

BENJAMIN

SIMEON

Jerusalem

ISSACHAR

ZEBULUN

Hebron

GAD Arad

Beersheba

☆ Tamar

☆ Kadesh

Elath

MEDITERRANEAN SEA

Jordan

SEA OF SALT

ARABAH

Brook of Egypt

0   50 km

© Michel van Elk, 1996

Borders are basically unstable because of the uneasy equilibrium of stationary growth based on the antithetic forces of political integration and economic differentiation. The idea of unstable borders is not new but is a basic notion in French geography. Gottmann (1952, 1-19), the eminent geographer who stands in the French tradition in which history and geography are conceived as two sides of cultural development, called the political *cloissonnement du monde* the result of an uneasy equilibrium between economic movement that ignores borders, and political relationships that are meant to protect people with their culture against others. The two forces are antithetic. Economic movement is opening up the world by creating nodal cities and networks and stimulating differentiation, whereas protective stability is compartmentalizing the world by drawing borders enclosing lands. The differences between Gottmann's views and those expressed here are that economic movement is an unjust process of ongoing differentiation creating all sorts of externality fields, and political protection is a just political action if it aims at protecting the land of the neighbor from negative externalities. In view of the antithetic character of economic differentiation and political integration, my thesis is that the disappearance or absence of borders is a sign of injustice, whereas their presence is a sign of justice. This thesis finds support in the biblical atlas, better known as Genesis 10-11.

The author of Genesis was very much concerned about differentiation and integration, which he saw as two antithetic forces shaping history. He distinguished three global models of human cohabitation: the *Island Model*, the *Babel Model*, and the *Hebrew Model*. The political differentiation of the world in its most extreme form is represented in the Island Model. It is the first model after the flood, so a world full of borders is the embodiment of the concept of regeneration after a period of injustice in which Noah was the last representative of a just individual (Gen. 6:8; 7:5). The Babel Model represents a world without borders as the ultimate form of political integration. It is the "historic" successor of the Island Model. So the cultural development in Genesis 10 is shown as a historic process from a regenerated, highly differentiated but hardly integrated earth (the Island Model) to a degenerated, highly integrated but totally homogeneous world-system (the Babel Model). In the Island Model each nation lives in its own country (island), everybody lives *according to a particular language* in his tribe and in his nation. In this model, each nation is a conglomerate of tribes united by a common language and distinguished from other nations by a language barrier. This extreme form of political and cultural separation depicts a just world and the beginning of a new curve of development. The Babel Model represents the end of the curve. In Genesis 10 this is the sequence of historic events. The Gottmannian world of relatively independent regions and nations developed into a Wallersteinian world system with Babylon as its capital and Mesopotamia as its core region. The rest is occupied periphery (the desert) and semiperiphery (Canaan, the fertile crescent). This time there is no natural catastrophe. It is the familiar human drama of fundamental misunderstanding between individuals who only hear their own voices that brings an end to a historic period of injustice, symbolized by human hubris, causing man to dethrone God by building a temple that reaches the sky for Baal, instead of smashing the idol. Genesis 11 tells us why political empires collapse. The lesson of Genesis 10-11 is that the presence of borders signals justice, whereas their absence is a sign of injustice. In the Hebrew Model, also shown in Genesis 10-11, borders are a theological necessity, dividing the world between Hebrews (migrants, people in search of land) and the peoples who have land. Theo-historically, the Hebrew Model represents the intermediate stage between deliverance (from exile in the land of a superpower) and regeneration (somewhere else). Geographically, the Hebrew Model represents the periphery, the region contested by two superpowers (Egypt on the Nile and Persia on the Euphrates, cf. Genesis 15:18). This geo-historical in-betweenness of the model is expressed in the term Hebrew signifying migrant and region beyond (the river). But the term Hebrew also signifies fury or rage. All three meanings are relevant in connection with the concept of encounter.

Hebrews are born in exile because their unrighteous fathers and mothers were exiled. As their children, these Hebrews are innocent to begin with. But how will they encounter the people who live on the other side of the border? As people overflowing with rage such as Simeon and Levi in Genesis 34? Or do they enter as Abram, inviting the inhabitants to make covenants with him and by doing so fulfill a fundamental human duty, which is to seek peace? Given his name, origin, and hometown, Abram decided to become a border crosser (Hebrew), covenant (alliance) maker (Heber) in the town of the ally (Hebron)! Regeneration after deliverance does not simply happen, it involves an encounter. In between deliverance and regeneration one is an alien in a foreign land because one cannot claim land on the basis of a past inheritance. The past does not legalize any claim on land; only righteous behavior in the present will secure the land for the future generation. In short, borders are a meaningful concept in connection with justice and encounter because they remind people on either side of their common history and the fact that this history does not legitimate claims, rights, or wars. This lesson is further elaborated in the two borders that Joshua described when he divided the land into eleven unequal regions.

Borders as Places of Encounter and Sources of Justice

In our society borders are made to divide, to separate, but in the biblical narrative it is the other way around: Borders join what is divided. The text on which the map of Joshua (Map 5–2) is based is an excellent illustration of the significance of borders as places of encounter. After dividing the fatherland according to the divine principle of inequality by incorporating recent history into his new geography of Israel, Joshua had to complete his model with a form of interregional integration that would keep the divisions together. He also had to incorporate a form of international integration that would unite the collection of peoples and their lands located between the Nile and the Euphrates. These two forms of integration, which safeguard an existing cultural differentiation, are found in the descriptions of two specific borders. Their stability is so important that the precise location of each of these borders is described four times.

The interregional boundary is the line between Joseph and Benjamin; the international boundary is the line between Judah and Edom (Hoekveld-Meijer 1996, 373-408). The Benjaminite-Josephite boundary runs all the way from Jericho through the Shechem Pass toward the Mediterranean Sea. This ancient international highway cuts the Shechem Pass area, the most contested crossroad region of Canaan, into two pieces. The significance of this brand-new border is twofold: It separates and unites. It separates Rachel's sons: Joseph, who represents the exiles who were born in Babylon (the capital of postexilic Aram), and Benjamin, who was born in Canaan after the exile. The border thus prevents either of the sons of Rachel from controlling the main crossroads of Canaan. The Aramaean Israelite Joseph is reborn as an Egyptian Israelite represented by his two sons Ephraim and Manasseh. Therefore, we can say that the border separates the *autochthonous* Benjaminites from the *allochthonous* Ephraim and Manasseh. Yet, the border is described as a thoroughfare and a perennial river full of water. Instead of separating allochtonous and autochthonous brothers, this interregional border seems to unify them. This curious border is marked by the court of justice that Joshua established in Shechem at the feet of Mount Ebal, and the Tent of Meeting that he established on Mount Gerizim opposite Shechem (Joshua 24). The autochthonous people have to cross the river and go to the other side if they seek justice. The allochthonous peoples have to cross the river to seek and serve YHWH on Mount Gerizim in the land of the autochthonous people. At this place, between high court and church, peoples encounter each other at religious festivals three times a year in order to enjoy the fruit of the land and commemorate their past deliverance, exile, and regeneration (Deuteronomy 16:11, 14, 17). This plural society, characterized by social and economic inequality and festive social meetings located on the common border between the court of justice (a sacred place for learning the Law) and the sacred place of cult and prayer, is the alternative of the rather postmodern ideal Solomonic society where each individual sits under his own fig tree and vine to enjoy the taxes he had to pay to his colonial, kingly priest in Jerusalem (1 Kings 4:22, 25).

The national border of Judah and Edom is more or less of the same quality. But instead of dividing and joining brothers, it divides and joins Israel and its archenemies. This curious border is also formed by a perennial river. It runs all the way from Kadesh (Sacred Place) via Beersheba to the sea. Kadesh itself is on the Edomite side opposite Arad (in Judah) and the place where the

Egyptian Israelites had the chance to regenerate but did not take advantage of it. The Edomite city Kadesh is also called En-mishpat or *source of justice* after the destruction of Amalek, the black sheep of Edom, by a worldpower (Genesis 14:7). Amalek is the proverbial enemy of the poor and the weak. Kadesh is the place of the altar that reminds people that YHWH is at war with people like Amalek who prevented refugees from entering the promised land. The place also reminds Israel that it must uphold the law in order to suppress injustice. Justice is personified as a trinity consisting of Moses (the Law), Hur (the army), and Aaron (the priest). As long as the army and the priest support Moses and his law of duties, the war against people who support social injustice will be successful. If they relax, injustice (Amalek) will win (Exodus 17).

The Edomite town Kadesh, the sacred place that served as a place of regeneration, is also the city where Moses encountered the second archenemy: the Ishmaelite Midianite Jethro (his excellency, his remnant). And once more, Israel must learn that one should not condemn a whole people because there will always be pious ones among the oppressors who know how to prepare the way for justice when the people have forgotten. On the border of Edom and Judah, Jethro taught Moses how to organize a society characterized by biblical justice (Exodus 18) before Moses received the Law (Exodus 19). At Beersheba, the other name marking this border, we meet Israel's third archenemy. This is the place where Hebrews and Palestinians made peace and where the Hebrew credo par excellence was formulated by Abram the Hebrew: *YHWH is El (or Allah) forever*, and the Philistine leaders recognized that El is with Abraham because Abram did not behave like a fury such as Levi or Simeon did in Shechem, the northern border town. Beersheba is the place of a theological revolution on both sides: It is the place where archenemies redefine their traditional concepts of God so that they may survive together in one and the same land.

In short, in the Old Testament, borders divide and join the two modes of human existence: They divide and join brothers of different origin as well as brothers or neighbors who turned into enemies. This metamorphosis is visible in Hebrew script where *r`* means *brother by birth or covenant*, and `*r* means *enemy*. So the kingdom of God is perceivable on two levels of analysis, thanks to its borders. On the level of the parts, God's own country is a plural society with internal borders. On the level of the whole, God's kingdom is a confederation of nations. Both kinds of borders are marked by national boundaries, by capitals, and by places of encounter where people are reminded of the past in order to maintain justice in the future. This means that, in theory, there are many countries that might be called God's own country. That is why this contribution to the discussion of geography and a Christian worldview is titled "God's own countries." The plural is unusual, but defensible if one knows that externality field and regional inequality, two concepts that imply the existence of borders, are meaningfully related to the theological concepts of regeneration, deliverance, redemption, justice, and encounter, as well as to past, present, and future.

Two problems remain. The first is how the traditional geographical relationship between humanity and nature relates to the Christian geography as proposed here. Or, how do we integrate the familiar concepts of stewardship and natural environment into this conceptual structure? The second problem is how to incorporate the gospel of the New Testament with its emphasis on individual salvation into this conceptual structure with its emphasis on theo- and geo-politics. I answer these questions in the order in which I raised them.

Stewardship and Nature in the Light of Justice

I draw attention to the function of nature in the biblical world picture. As stated before, in the Old Testament, nature is not an autonomous god or conglomeration of gods intervening in the lives of people in an utterly arbitrary way. Nature has the modest function of *hand of God*. It blesses the people, when they maintain justice (with rain at the proper time); it curses the people with catastrophes such as locust, drought, and earthquakes when they ignore God's commandments (Houtman 1982, 50–59). The heavens and earth are assigned by God to serve as witnesses to peoples' behavior in the context of the covenant between a people and their God (Deuteronomy 4:26; 30:19; 31:28; 32:1; Houtman 1982, 76-77). Sun and moon are assigned to be the mechanics who must maintain the natural cycle in order to maintain life on earth (80–82). In other words, nature is not a phenomenon that ought to be treated with respect as among environmentalists (and in nature religions). *Nature is part of the biblical juridical system.* God is judge, and nature is witness and instrument. If cultural development is wrongly directed and people are not willing to restore justice by taking measures as indicated in Figure 5–1c, nature will intervene (by stopping the environmental cycle or other events) in order to transform the destiny of people from the blessed status of stationary justice to a temporary finale.

The concept of stewardship (Curry-Roper, above, chap. 4) hardly fits into this biblical concept of nature. But stewardship that does not necessarily imply a subservient relationship to nature is not at all in line with that of *dominium terrae* formulated in Genesis 1:26, 28, nor is it in line with the function of the accountant in the New Testament (Matthew 20:8; Luke 8:3; 16:1ff. Romans 16:23). The biblical author of Genesis used a verb to describe the nature of *dominium terrae* that indicates suppression by force, a trampling down (*kabash, radah*). It is the semantics of this verb that has caused a lot of misunderstanding that, in turn, has such a negative influence on humanity's attitude toward nature (the famous thesis of White [1969] about the negative influence of Judaeo-Christian thinking about nature). People interpreted the verb without knowing its sociohistorical context. We must realize that the first chapters of Genesis are meant as antidote against the Babylonian and Egyptian world pictures. In these world pictures the fate of man is arbitrarily decided by the gods of heavens—the masters of the earth (in the Babylonian creation legends)—and ruled by animal-gods (Egyptian mythology, Baal religion). The concept of *dominium terrae* reverses the roles (cf. Westermann 1974, 219). Nature is not a god who rules the earth. People rule the earth. Nature does not rule but responds to the demand of YHWH, depending on whether the interrelationships among peoples are just or not. From the biblical perspective of *dominium terrae*, environmentalists should know that nature is best served when people restore social justice.

When we demythologize the concepts of stewardship and natural environment, these have a place in the conceptual structure represented in Figure 5–2 as a possible mix of imagination and conceptuality that may help Christian geographers to establish a branch of geography in which the themes of redemption, encounter, regeneration, justice, and deliverance are an integral part of the theoretical structure.

Figure 5–2
A Christian worldview: A mix of imagination and conceptuality

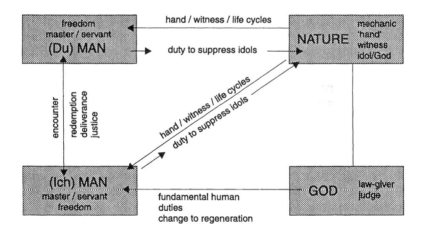

Redemption and Deliverance in Christian Geography

The universal connotation of the Christian concept of the kingdom of God seems incompatible with a politically divided world. People are equal in the sense that they can be delivered and redeemed. So, the last question that needs to be answered is how to relate the theme of personal redemption and salvation, first preached by the apostle Paul, to that of collective responsibility for social justice in a politically divided world.

Luther already discovered that Paul's gospel is not in the gospels of the evangelists. They "do not speak properly about the death of Jesus Christ" (cf. Schenk 1983, 12-13, quoting M. Luther, 21, 223). However, the contrast between the gospel of Paul and the other gospels can be explained. This explanation is relevant to our question and to the general theme of the chapter. So, let me expose the historical contexts. The political division of the world ruled by the Roman emperors had no relevance for Paul because he believed that the Empire's end was near. But the evangelists, writing just before or just after the destruction of Jerusalem in 70 AD by the Roman emperor Titus, had to reinterpret Paul's letters written some ten to twenty years before the fall of Jerusalem. The day of judgment had indeed come, but was not accompanied by the physical return of Jesus Christ and a collective ascension to heaven (1 Thess. 4). Hence, Mark, followed by Matthew and Luke, gave another description of the gospel. The first lines of the gospel according to Mark are found in Isaiah 40. This chapter in Mark refers to a political situation some one and a half century before Isaiah wrote his prophecy. The political situation described in the chapters preceding Isaiah 40 is similar to that of Mark. Jerusalem is beleaguered by a foreign emperor. But the king, the representative of a collective, repents and the army disappears in the night. In Mark 1:14-15, Jesus urged his public to "repent, and believe the gospel." Given the historical context of these words and the univocal reference to Isaiah, Jesus says that Jerusalem will be delivered if its leaders repent. But is not enough that the leaders repent. Referring to Isaiah, John the Baptist urged the people to restore social justice in order to prepare the kingdom of God in Judaen society, just as Isaiah urged his people in his days. They should level the hills—places where Baal the god of fertility used to be worshipped in the days of Isaiah and fill up the valleys—places where children were sacrificed in order to satisfy personal needs. In Luke 6, Jesus reads and explains the gospel of Isaiah 61. The prophet predicts that the foreign emperor will be destroyed and Jerusalem rebuilt. This will happen when the eyes of the blind are opened and so forth. Luke testifies that Jesus opened the eyes of the blind and made paralyzed people walk again. Given this sign, Rome will fall, not by a revolution but by the hand of God who will send a new world power to destroy it. These two references to Isaiah demonstrate that a politically divided world has a theological purpose in the gospel of Jesus. There must be a politically strong neighbor to deliver his neighbors from injustice. That is why the theme of collective redemption and deliverance has a theo- and geo-political meaning not only in the Old Testament but also in the gospels of Mark, Matthew, and Luke.

Epilogue

When we take the conceptual structure of Figure 5–2 as a meaningful compass, it is clear that we cannot isolate Christian geography from other disciplines. It is interrelated for example with the Christian variant of history and law. Moreover, it is applicable to Christian regional and city planning and to geographic education. By a *Christian variant of history*, I mean that kind of history in which the past and present are studied from various points in the future, and time is perceived as a sequence of mutually related generations. When necessary, these relationships need to be expressed in terms of guilt. In these historical studies, law, formulated in terms of fundamental human duties, is the hermeneutical key to interpret the past and to reach an understanding of the present. When history incorporates this triple perspective, it is a meaningful guide for regional planning and environmental studies. By *Christian regional planning* I mean the variant in which goals are formulated in terms of regeneration, restoration, and compensation and not in terms of accommodating future growth. By law I do not mean a Christian code of conduct but, rather, fundamental human duties.

On the basis of my construction of a Christian worldview, I suggest the following definition of Christian geography, paraphrasing the definition of the U.S. National Geography Standards of 1994: *Christian geography is the study of regional inequality, externality fields and borders, always from the perspective of the recent past, the present and the future of one people to all others.* The goal of Christian geography may be formulated as follows: *Everyone who studies Christian geography must possess the knowledge and the skills to analyze political and economic relationships that are caused by or cause externality fields and unjust regional inequality, (including those physical processes that cross borders), and the knowledge to interpret such an analysis in order to exercise one of their fundamental human duties: to contribute to the elimination and restoration of negative spillover effects, and the maintainance or restoration of a just regional inequality.*

Notes

[1] The reconstruction of the map of Joshua is based on an analysis of the semantics of toponyms mentioned in the detailed border descriptions in the book of Joshua (Hoekveld-Meijer 1996, Appendix A and B). Joshua drew a blueprint of the promised land for the Israelites who returned from Egypt; Ezekiel drew his map for the Israelites who planned to return from Babylon. There is a growing consensus to consider Genesis –2 Kings as a postexilic work (Houtman 1986, 21; 1993, 1). If it is such a late work, the maps of Joshua and Ezekiel are contemporeneous and could be considered as alternatives of how to divide the land in a righteous way. (For a late provenance of historical bibliography, see for instance, Clines 1978, Wesselius 1995). For the idea of an invented history of Israel, see Edelman 1991, Bretller 1995, Whitelam 1996. For a late exodus and a late Moses, see Van Seters 1994. For the semantics of names, see for instance Garsiel 1987, 1991; for the geography in names, see for instance Deurloo 1990, Engelken 1990.

Chapter 6

Alien in a Foreign Land: Human Geography from the Perspective of Christian Citizenship[1]

Gerard A. Hoekveld

> *Thus you (inhabitants of Ephesus) are no longer aliens in a foreign land, but fellow citizens with God's people (the Jews), members of God's household* (Paul's letter to the Ephesians 2:19).

Introduction

THE AIM OF this chapter is to stimulate discussion about the focus of human geography from the perspective of Christian citizenship. I used to say that geography might be good or not so good but not Christian, Buddhist, Atheist, and so forth. Geography has its own conceptual structures. At more concrete levels, these are embodied in theories and descriptions of facts and relationships that are disclosed by geographical craftsmanship. I thought that ethics, or other normative issues, inspired by any religious belief or philosophy, comes into the picture when geographical knowledge is applied. In other words, ethics enters at the end of the empirical cycle; it does not stand at the cradle of knowledge. Now, I am no longer convinced of this. I submit that Christian geography is more scientific than often assumed if it succeeds in integrating the concepts of *citizenship* and *externalities* into the conceptual structures of geography and their empirical concretizations.

I argue that these concepts are also basic to geographic education in Christian secondary schools. It is not possible to consider geographic education without taking into account geographic theory in general. Geography teachers are dependent on the results of academic geography. They select and transform these in keeping with the aims and constraints of geographic education in the schools. If existing theories do not fit the aims and goals of school geography, then academic geographers may need to reexamine and overhaul them. We should turn first to academic geography in order to find ways to examine its stock of theories and empirical knowledge. In this contribution, the concepts of citizenship and externalities will be reviewed so that they can serve as keys for such an examination.

Why Citizenship and Externalities?

Citizenship is a fashionable topic today (Turner 1990, 189). It is treated extensively in political science, geography, and social-science courses because of its moral dimension (see, for example, Williams 1975; Smith 1989, 147), often in the context of problem solving (Livingstone 1992, 216). For example, the U.S. National Geography Standards 1994 (National Geographic Society, Research and Exploration 1994, 28-29) express a commitment to cultivate geographically informed persons who understand "that geography is the study of peoples, places and environments from a spatial perspective and who appreciate the interdependent worlds in which we all live." This aim is framed by the 1990 National Education goals that state: "By the year 2000, every adult American will be literate and will possess the knowledge and skills to compete in a global economy and exercise the rights and responsibilities of citizenship . . . " (244). Nevertheless, in many school and university courses, the moral components of geography are not "very enhanced by the largely technical interpretations of the physical and moral world" (Smith 1995, 272). Of course there have been many geographers who show a strong ethical component in their work. Well-known examples are the books of D. M. Smith (1979, *Where the Grass Is Greener: Living in an Unequal World*) and D. Harvey (1973, *Social Justice and the City*). Moreover, many geographers practice applied geography because of their ethical concern about the state of the world. Still, Smith is right. When there is no explicit ideological basis, like Harvey's (1973, 15), ethics mostly remains implicit. To some degree, this situation is caused by the increasing pluralism in our society—a condition that is often equated with a moral pluralism. The state and the agencies that have to specify aims, goals, and standards for geographical education then have to resort to a technical approach in order to avoid endless debate about philosophy, society, and worldviews. Yet this attitude reinforces the dividing lines separating people of different moral outlooks to the point that each is an alien in the eyes of the other. In such a context, citizenship is an empty concept. But in spite of the moral fragmentation of society, people are related to one another through their political participation and through so-called externalities. This brings us to the geographical dimension of ethics.

Despite the differences among geography's traditions, such as Marxist or mainstream geography, there is basic agreement on human geography's core concept: spatial organization. Although it is ill-defined, the concept implies that cohabitation is one central focus of human geography's interest (Gottmann 1952, 192). Cohabitation denotes that people share the same area by their contiguous locations and spatial interaction (155). Spatial organization can be analytically described as the whole complex of land use—ownership and control relationships between elements of society (e.g., firms, households, governmental agencies), the parts of the earth's surface that together can form an area and the conceptualizations of the land that form the basis of these relationships (Hoekveld 1990, 16). Those relationships have effects upon the territory, the interests that maintain them, and the society as a whole. But spatial organization generates causes and effects outside its territory. Such effects are externalities; the area in which these effects are experienced is the "externality field" (Harvey 1978, 281) or simply, the locational and spatial context (Cox 1972, 11). If externalities (or spillover or third-party effects) are perceived as positive by one of the parties in the externality field, that party will try to internalize them by erecting fences or border posts or simply by drawing lines on a map. If the effects are experienced as a nuisance or a threat, the party will try to externalize them. In either case, externalities are a basic mechanism to create conceptually and legally bounded spaces (Harvey 1978, 282). Figure 6–1 depicts the relationship between spatial organization and externalities. At the same time, it shows the ethical imperative of spatial organization. It is not enough to perceive, evaluate, and use one's territorial potential. One must also know which externality effects will arise, given a specific land use on either side of the border. Decision making must be oriented toward mimizing the negative externalities on the territory of the other. Good citizenship thus seems to imply reducing negative externalities on the territory of one's neighbor. In other words, citizenship is neighborship made explicit in a policy of "not in your backyard."

Figure 6–1
Externalities as effect or as ethical imperative of spatial organization

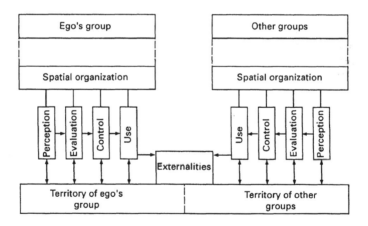

Whereas conceptually bounded spaces are the raison d'être of human geography in general, legally bounded spaces are the specialty of regional and political geography in particular. It was in legally bounded space that citizenship emerged and developed. Hence, citizenship and externalities are a conceptual pair best studied in the context of political and regional geography. This means that the definition of citizenship should be couched in regional and political geographical terms. However, citizenship can also be defined in Christian terminology, where it functions in the context of the kingdom of God—a place where people are no longer aliens or political neighbors but fellow citizens. In the following section, I shall prepare the ground for a Christian regional geographical definition of citizenship.

Citizenship as Regional Geographical Concept

Although citizenship is quite a familiar term, it is often ill-defined if defined at all. It may refer to the *status* of an individual, e.g., the duties and rights of the members of a specific community, be it a village or a country or even the world. A citizen is one "who counts with respect to certain obligations and rights" (Smith 1995, 275). The term may also refer to the behavior that individuals display or are expected to show with regard to the polity of which they are a member. And the term may denote the attitudes that the individual should have. Those attitudes and the behavior they engender are called *virtues* because they contribute to the well-being of the community and its members. Heater (1990, 193) even calls "the ideal good citizen a paragon of virtues who brings to the fore different qualities according to the circumstances." In other words, virtues are at the root of positive externalities. Educational aims such as the personal development of individuals, the preparation of the student for later life as a wage earner making a decent living in a capitalistic society, the preparation for later studies, and the education for participation in society—all promote virtues within the context of citizenship. Particularly, the U.S. National Educational Goals concerning the uses of geography (e.g., standard 18, grades 9-12) refer to the aims of citizenship. The objective of the American National Geography Standards is "to develop world-class levels of understanding of geography which will be useful in the context of work-place, voter's booth and people's lives in the United States" (National Geographic Research and Exploration 1994, 253). In the same standards, answering geographical questions is viewed as "a skill linked closely to good citizenship" (44). The practical application of this skill "needs to be fostered in all students in preparation for life as the responsible citizens and leaders of tomorrow" (103). Status and virtues are attributes of individuals, but at the same time these are relational attributes that link individuals to specific communities and bounded spaces at various scales. The definitions of citizenship that are common in political philosophy, however, hardly ever relate an individual's status and virtues to the various scales of bounded space. Exceptions include, for instance, Heater (1990) and Selznick (1992). The latter states that "when the community expands particularism is diluted" (Selznick 1992, 195).

Political philosophy is focused on complex relationships between the individual and the community without taking externality effects into consideration. Because the community does not cover the whole world, the ignorance of these effects is a conceptual shortcoming in the theory of citizenship. I summarize the debate in order to define the regional geographical dimension of citizenship.

Rawls (1971) represents the view in which the state is a partner. He formulated in his "theory of justice" a new form of liberalism, called "social liberalism." His point of departure was that the state should safeguard the liberty and equality of all individuals. The state should correct the mechanisms of capitalism that threaten those principles. Furthermore, it should maintain a social order in which all individuals can realize their understanding of the ethically good life regardless of religion, talents, or ideas. According to Rawls, the outcome of the self-interest of all individuals creates an idea of justice. This entails that all social goods, freedoms, opportunities, income, welfare, and all other conditions that create self-respect should be distributed among all individuals in an equal way, unless an unequal distribution is advantageous for the most needy (Rawls 1971, 154). This idea is based on a very individualized understanding of ethics, namely, individuals' self-interest and their instrumentalist acceptance of the constitution that specifies their rights and duties.

Rawls' theory of justice is not very practical in conflicts arising from externalities in the externality fields. The simple reason is that negative goods (such as pollution), or positive goods (such as upstream river flow regulation) are subject to distance decay and do not stop at national boundaries. The so-called communitarians attacked Rawls' social liberalism. Their critique contends that individuals are not isolated from, but are always part of, wider communities. So, one's ideas are at least partly determined by shared perceptions of the ethically good life. Moreover, the community is the basis for the continued existence of personal liberties that could not survive under the pressure of self-interest. Communitarians fear a bureaucratization and juridification of the public sphere and a withdrawal of citizens into the private sphere if the social liberals were to have their way. At present, the positions of both camps are converging. Both positions were one-sided. And a liberal society is dependent on concrete support by the populace of the ideals of freedom and equality (Van den Brink 1994, 22). What communitarians and social libertarians have in common is a blindness to the externality effects of spatial organization beyond the boundaries of community and state. Kymlicka and Norman (1994), for instance, distinguished four different positions in the debate, superseding the liberal and communitarian positions). First, the "new right," accuses the welfare state of creating passive consumers of rights, of entrenching bureaucracy, and of losing civic virtues. Second, the "participatory democrats," will teach people tolerance and responsibility as prime civic virtues. Some of them, known as the "civic republicans," even propagate political life as the highest form of being human. Third, the "civil society theorists," starting from a communitarian position, think that civil society itself is a seedbed of civic virtue. In voluntary societal organizations such as churches, trade unions, and so forth, civic virtues are learned. Fourth, the "liberal virtue theorists" believe that schools should be responsible for teaching civic virtues, especially a willingness to engage in public discourse and to question authority.

T. H. Marshall, who started the ongoing British debate on citizenship in 1950, recognized the effects of social segmentation on social justice. Yet he did not recognize the externality effects that derive from spatial organization. His main concern was the tensions between a capitalist economy, seen as causing social inequality, and the welfare state, seen as a condition for citizenship based on political equality. He distinguished three types of social rights and their corresponding arenas: civil, political, and social rights with their respective institutions of law courts, parliament, and the welfare system. Marshall defined these tensions in terms of class struggle. Turner (1990) and Garcia (1996) put the ensuing debate into perspective by noting the element of regionalism in these philosophical discussions. Noting the parochial British context of Marshall's analyses, both scholars pointed out that the ideas about citizenship differed strongly among European countries and that they are still evolving. In Turner's opinion, the dimensions of public versus private domains, of homogeneity versus ethnic pluralism, and of citizenship based on national versus global membership have been neglected (Turner 1990, 212). Attention to these dimensions would lead to a political philosophy in which citizenship operates on a global scale. At that scale, boundaries of any kind are irrelevant, and externalities are de facto nonexistent.

Habermas (1992, 1) takes the view that the world is the ultimate scale of citizenship. He analyzed the conflict between the universalistic principles of constitutional democracies, on the one hand, and the particularistic claims of communities to preserve the integrity of their habitual ways of life, on the other. He noted a development of the concept of citizenship away from a status in national and international law that assigns a particular person to a particular nation whose existence is recognized in terms of that law. Instead, citizenship moved toward a status of membership in a state defined by civil rights. Modern developments draw nations into a single communication system. The result is a continuum with state citizenship at the one end and world citizenship at the other. The development of a European constitutional patriotism might be the first step in the direction of such world citizenship. In the vision of Habermas, the world will be without aliens in a foreign land when societies can supersede different interpretations of the same universalist rights and constitutional principles now still marked by the context of different national histories (cf. Habermas 1992, 12). In other words, Habermas' vision does not differ much from that of the apostle Paul as described in his letter to the Ephesians, in which God's household eliminated ethnic boundaries (Eph. 2:19). Today the universal concept of God's household has been replaced by universal ethics dominating cosmopolitan or world citizenship—at least on paper. In the United States, Martha Nussbaum (1995) has pleaded eloquently for a cosmopolitan citizenship that pushes back patriotism as the main source of national citizenship. She concedes that such citizenship results in the loneliness of the citizens if not compensated by a universal love for humanity and by universal reason (28). The reaction in America has been unfavorable for the most part. Her ideas are seen as unrealistic. Moreover, they downplay the importance of patriotism in shaping personal and community identity. Regional geographers may add that ideas like this downplay the externalities that tend to strengthen patriotism. This is especially the case when the externalities are experienced as negative by the other party in the externality field.

The idea that patriotism is shaped by duality—as in the case of externalities marking inside and outside—is recognized by Waltzer (1994) in the field of ethics. In his opinion, duality of universalism and particularism in a culture is a necessary and natural concomitant in the formation of communities. He distinguishes a "thin morality," which is minimal morality that is universal and present in all cultures. It is embedded in a "thick morality," which is expressed in a specific idiom shared by people within one culture having the same historical, cultural, religious, and political orientations. Without saying it in so many words, Waltzer thus accepts the phenomeon of a regional moral order as a specification of a universal one. The thin morality seems to come down to truth and justice but requires a context of thick morality within which it is concretized. The philosophical bedding of thick morality is contextual pluralism—one of the three forms of pluralism analyzed by Griffioen (1995, 216-220). In this form of pluralism, contexts are considered to be unique. Hence, they require specific applications of universal rules. This is called situational ethics by many Christian philosophers and theologians.

Thick morality thus mitigates the rigid application of universal ethics but does not relativize their inherent claims to truth (Griffioen 1995, 217). Given its emphasis on context—always in time (history) and place (geography)—contextual pluralism justifies taking thick morality or regional ethics as the ethical dimension of studies in regional geography. In that light, regional ethics should be studied from the perspective of universal ethics. Then the comparison of externalities in a regional geographical context would not be restricted to an evaluation of spatial configurations and processes. Rather, that exercise should be preceded or followed by a comparison of the regional ethics in question in order to judge the spatial patterns. This is nothing new. In her study of theo-political relationships between Edom and Israel, Hoekveld-Meijer (1996) demonstrated that the historical books of the Old Testament contain a dialectic casuistic about good and bad leadership. An appreciation of the geo-historical context of the biblical narrative and law is necessary to interpret the regional ethics.

At this juncture I return to the task I set out for this chapter, though not by the shortest route. On the one hand, I am in search of a regional geographical definition of citizenship in relation to the status and virtues of the individual within the context of thick morality, and, on the other hand, I approach externalities as spatial context of aliens in particular. In this regard I should like to take note of similar efforts along these lines. Let us consider an example of conflicting thick moralities. In his comparative study of *Citizenship and Nationhood in France and Germany*, Brubaker (1992) describes two complementary applications of universal ethics in two different geographical contexts. That example may help me to better formulate the definition. Brubaker introduced the cultural dimension that relativizes the concepts of nationhood and citizenship as these are used in the general debates among political philosophers. He demonstrated that understanding the perception of a given political body is essential for the concretization of rights and duties and the selection of expected virtues of the citizens. This self-understanding of nationhood, national identity, or a nation-state is the outcome of historical processes, mostly of long duration. According to Brubaker, citizenship is the concrete manifestation of a person's status with regard to the state to which he or she belongs.

> Citizenship refers not only to political rights but also to the unconditional right to enter and reside in the country, complete access to the labor market and eligibility for the full range of welfare benefits. In a world structured by enormous and increasing inequalities between states in labor and consumer markets, welfare systems and public goods such as order, security and environmental quality, the rights conferred by citizenship decisively shape life chances. The question of access to the territories, labor markets and welfare systems of the world's favored states is decisive for persons and states alike (Brubaker 1992, 180-81).

Obviously, access to public goods, to which citizens have a right, is an extremely negative externality for aliens.

Brubaker distinguished two determinants of citizenship. One is the *ius soli*, which ascribes citizenship to place of birth and residence (for a given duration). The other is the *ius sanguinis*, which ascribes citizenship to genealogical descent, independent of place of residence or birth. The *ius soli* determines French citizenship. There are no second generation aliens. German citizenship, on the other hand, is based on the *ius sanguinis*. Accordingly, aliens are defined differently in these two juridical systems. It is clear that these differences will be apparent in the selection or emphasis on particular civic virtues. The definitions of aliens are the same in the sense that differences can be deduced from one-and-the-same expression of universal (thin) ethics: It is not good that the man should be alone (cf. Genesis 2:18). People form social groups. They are either citizens by birth in a land in which citizenship is determined by the *ius sanguinis*, or they are citizens by free will.

In the latter case, people decide to stay in the country of their parents with its *ius soli* or to migrate to such a country. A society based on the *ius sanguinis* is a closed society but has no bounded space; a society based on the *ius soli* is an open (plural) society within a bounded space. The philosophy of Blut-und-Boden is an ethical monstrosity because it combines the two principles, creating a closed society in a closed space. The biblical society as described in the Old Testament is based on the *ius soli* because the people are bound by covenant not by birth. Arguably, the *ius soli* is preferable from a Christian point of view. Community is not a matter of birth but of divine law (Matt. 12:46–50; Mark 3:31–35; Luke 8:19–21). This would justify defining citizenship in terms of the *ius soli*, which is based on humankind-partitioned political boundaries. In a world thus divided, citizens in one region are perceived as aliens by citizens in other regions; they experience each other's externalities. Therefore, I propose the following definition of citizenship: Citizenship is the legal status of individuals by which people are committed to a particular territory and to each other.

From an ethical point of view, this definition is not complete. It does not mention that people are responsible for the negative externalities experienced by aliens. Hence, I would like to expand the scope of commitment as found in the definition by introducing the concept of compensation. This is one of the key principles on which communitarian justice is based (see Selznick 1992, 431-32). As a central concept, compensation is derived from the *regula aurea*, which belongs to the thin morality of Waltzer and is also a principle of biblical morals (Perkins 1992, 657). The *regula aurea* is the command: "to love your neighbor as yourself" (Matt. 22:39, Mark 12:31). This command is a universal command (see, for instance, Stauffer 1959, 55-60, who showed the universal character of the *regula aurea* in a series of examples collected from all over the world. The commandment to love is also the quintessence and criterion for the interpretation of the Torah (Schrage 1982, 444, Smend 1982, 432). Jesus applied the Law to the fullest extent (Schrage 1982, 438, Schnackenburg 1986, 73). If love for the other is the quintessence, it may be expressed by compensating the others for any injustice inflicted on them by the one who loves them! Citizens should consider others as if they were their fellow citizens and act accordingly with respect to negative externalities. Although the others are, per definition, beyond their borders or outside their control, citizens must compensate the others when they are affected by negative externalities. Citizens should not attempt to integrate externalities by moving borders when the externalities are positive. If they treat others as their fellows, they are good citizens in an ethical sense. With this in mind, I can also formulate a definition of good citizenship that incorporates this basic civic virtue: Good citizenship is the legal status of individuals who are committed to a particular territory and to each other and who compensate negative externalities inflicted upon people beyond their borders.

The second definition seems to lead the way to a regional geography based on universal and situational ethics. It leads away from the concept of distributional ethics connected with the universals mentioned by Smith (1994, 293). Those universals are the capacity for pleasure and pain, some basic needs, and the necessity for a place in the world. Smith describes these in an individualized, egalitarian way, emphasizing the material aspects of well-being. My definition also answers Harvey's problem with thick morality and how to judge the traits of capitalism that differ in time and place. His solution is to confront the "faces of oppression" such as exploitation, marginalization, powerlessness, cultural imperialism, violence, and ecological harm with social policy and planning (Harvey 1992, 598-600). In my opinion, good citizens do not in the first place confront "capitalism's revolutionary dynamic which transforms, disrupts, deconstructs, and reconstructs ways of living, working, relating to each other and to the environment" (600). Good citizens should uphold the law in their own communities (regions, countries) and compensate the negative effects experienced by their neighbors. In this context, it is good to remember Heraclites' metaphor for good citizenship. He described good citizens as those who "fight for their laws as for their walls." (quoted by Selznick 1992, 434). Obviously, fighting for the law is not the same as simply applying a law, dependent on thick morality (often in the rigid form of legalism and the strict procedures of positive law). Upholding the law is putting thick morality to the test by thin morality. If thick morality stands the test, then it must be applied. If not, it should be changed. Hence, good citizenship is more than articulating and implementing "environmental justice" as Aay argued already twenty years ago. He proposed that the concepts of citizenship and environmental justice be incorporated into geography (Aay 1976, 19). Good citizenship is also more than "an ecological lifestyle" as was suggested by the managing editor of the *Christian Educators Journal* in a comment on Aay's paper (Aay 1977, 11). If one incorporates the concept of good citizenship to the discipline, then geography is more than a study of the mutual interaction between the cultural and the natural world. It is a study of geo-historical contexts and thick moralities evaluated from the perspective of universal law. In other words, in this kind of geography thin morality is the necessary third dimension of time and space (see Figure 6–2).

Good citizenship, therefore, implies examination and judgment of spatial organization within one's own city, region, or country as well as international spatial organization. Externality effects may occur on every scale.

With the definition of citizenship that is used in this chapter, it is clear that people who are not committed to each other within one state territory are not citizens in a true sense, although they are inhabitants with certain rights and obligations. For example, the behavior of people in Northern Ireland or the former Yugoslavia demonstrates that they are not so much citizens as merely inhabitants of their countries despite their administratively legitimated status. The concept of commitment of citizens to each other is formalized by the shared acceptance of laws of their state. In a plural society, commitment is institutionalized through laws that minimize or prohibit the negative effects created by one group and experienced by other groups.

If there is no acceptance of other thick moralities in a plural society, the externalities of both groups will create conflicts. In these cases, a very meager form of citizenship remains: the acceptance by both or all parties of procedural justice and laws. These are the rules and procedures that regulate contacts, cooperation, and conflict. In order to avoid an association with legalism, Selznick (1992, 331) prefers the term *process*. This process or these rules and procedures must be in correspondence with thin morality and will take shape in response to specific situations. Whatever a situation may be and

Figure 6–2
The time-space context of "thick" and "thin" morality in human geography

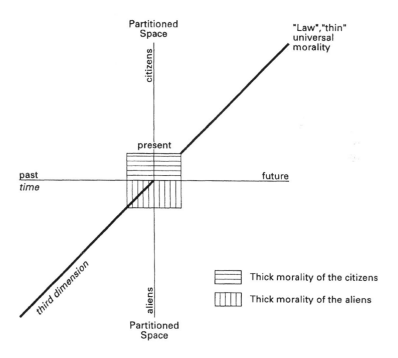

whatever term one prefers, every citizen is an in-law of all other citizens, irrespective of whether they are inhabitants of a uniform or plural society. Hence, in a regional geography that is based on such key concepts as citizenship and externalities, human beings are studied in their capacity as in-laws and neighbors respectively. In this kind of regional geography, ethics is at the beginning of the empirical cycle, not at the end. The same is true of political geography's investigating if a political or a social organization supports or undermines commitment to a territory and its people as well as to the people on the other side of the borders via externality fields.

In-laws and neighbors are well-known literary metaphors of aliens in biblical narratives about two types of responsibilities: the responsbility of citizens to one's household that includes aliens and the responsibility for the aliens in a foreign land—the land that extends beyond the fence of one's backyard.

Figure 6–3
The geographical framework of areal relations

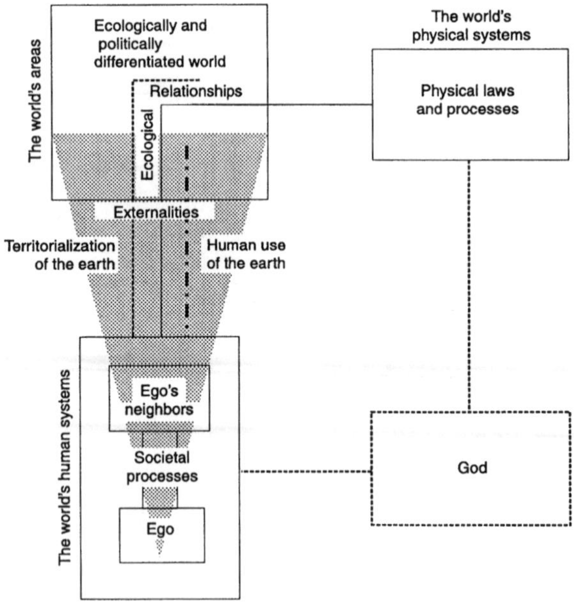

Figure 6–3 incorporates the more familiar geographic terminology used in the National Geography Standards of 1994 to depict the geo-political context of political and social institutions and organizations. In *Geography for Life* (1994, 33), the substance of geography is defined in terms of physical systems, human systems, places and regions (the "basic units"), and the relationships between these systems and areas. I have combined these "building blocks" of geography with the four elements of a Christian worldview depicted in Figure 6–1, page 85. The spatial organization of the

world's areas is classified in terms of human use and territorial division of the earth. The sociopolitical dimension of the world's human systems is represented as the system of ego and neighbors. Physical laws and processes are incorporated to underline the physical dimension of the world's area in relation to land use. Externalities occupy the whole orbit between ego and the ecologically and politically differentiated world. In this figure, God corresponds with biblical law. This represents the third dimension or thin morality in Figure 6–2, page 93. The question is whether biblical law can be seen as a variant of universal law. In other words, in which respect does the kingdom of God—a very central concept in Jesus' teachings—differ from the non-Christian utopias?

Kingdom of God, Divine Justice, and Regional Geography

Despite the distinction made between the Old and the New Testaments, we have to realize that the Old Testament was the starting point and the context of Jesus' teachings (Perkins 1992, 658). He did not abolish the Law of the Old Testament but summarized and fulfilled it (see, for example Matt. 5:18; Luke 16:17). The Law was maintained by Jesus as the expression of the Will of God, the Lord of His Creation (Keck 1996, 15). As such, it is the authority for implementing righteousness in the earthly part of His kingdom (Kremers 1962, 553). In other words, in the kingdom of God, God is King and law-giver; its citizens are those who commit themselves to His law (Walther 1990, 312). In that sense, there is no difference between the kingdom of God and, for instance, Habermas' universe. The difference must be in the law other than the *regula aurea*. But is there a difference?

The basic principle of this golden rule is reversal: Citizens are aliens, depending on the perspective in a particular bounded space. This reversal is best illustrated in the Bible quotation found at the beginning of this article. Paul tells the citizens of Ephesus that they are aliens in the kingdom of God and the aliens in Ephesus (the Jews in the diaspora) that they are citizens in the kingdom of God. But, at the same time, he removes the criterion that separates the two groups in Ephesus. The Ephesians and the Jews are fellow citizens. The reason why Paul removes the line of separation is a particularly rigid interpretation of the law by the Jews who consider themselves citizens of God's kingdom because they practice circumcision and observe laws with regard to their daily lives (eating and drinking customs). These are not ethical laws but laws that safeguard cultural identity through outward signs such as circumcision, although originally circumcision was merely a sign of a covenant. Paul simply removes laws that originated in another historical context or thick morality by replacing the sign of the covenant. It had lost is contractual character and had become a kind of birth certificate and ethnic marker. In other words, Paul recognized that the divine law had to be adjusted because the old sign had lost its original legal function. It had turned into a blindfold, preventing citizens from recognizing aliens as fellow citizens.

However, keeping covenants, whether divine or not, belongs to the domain of universal ethics. So what is the difference between divine law and universal ethics? And if there is a difference, what is its relevance for Christian geography in research and education? The concept of justice may lead us to an answer.

The term justice is shared by geographers, theologians and political philosophers, although their definitions differ. According to Finkel (1952, 411), the term justice is a semantic bridge between the intrapersonal and the interpersonal domains. In the interpersonal domain, the concept initially related to the maintenance of social order through laws and the administration of justice. "Justice means the outer limits of protection or rehabilitation on the part of the judge. The just judgment announces ways to restore harmony in the community, the social ideal of peace . . . " (translation mine). Jesus made justice the keyword in his preaching (Lührmann 1984, 415). Selznick (1992, 435) listed a number of mechanisms to restore harmony according to his theory of justice. He too argues that ethics must be grounded in historical experience. In other words, he calls for a thick morality and contextual pluralism. The mechanisms or virtues that restore harmony are entitlement (claims of rights granted by law), justification (in case of deprivations, the other side of compensation), equality in treatment, impartiality, proportionality, reciprocity, rectification, desert (merit), and participation. The idea of equilibrium unites these mechanisms. This is also the basis of the ideals of the French Revolution: All people are equal, and no one is more equal than any other.

However, in biblical law, the losers and the weak are more equal than others. God is a divine example of partiality in taking sides with the losers and the poor (in both the Old and New Testament). A situation in which losers exist is characterized by inequality. This inequality is compensated by a new inequality: The last are now first, and the rejected brother receives a double share (of land, cf. Hoekveld-Meijer above, chap. 5). With the concept of merit, it is the other way round. In universal ethics, reward is proportionate to merit. In divine ethics, reward does not depend on merit. In other words, the difference between divine law and universal law (a variant and a reversal) is that the former is based on inequality and unequal distribution, and the latter on equality and equal distribution. In universal law, justice prevails when right and wrong as well as merit and reward are in balance. In divine law, justice is reached when wrong is rectified by overcompensation and reward is independent of merit. The law of overcompensation as biblical virtue is most clearly formulated in Deuteronomy 21:15-17 and Matthew 5:38-48.

In regional geography, differentiation is conceived as a self-evident state of the world and is explained by all sorts of scientific models. If regional differentiation is in accordance with theoretic models, it is regarded as natural and logical; however, such differentiation is not natural in universal law, which aims at equality. In biblical law, regional differentiation is natural, though it is not acceptable unless it is the consequence of rectification and overcompensation. To put it in geographical jargon: Regional differentiation and unequal distribution are acceptable if they are the result of spatial organization that overcompensates for the negative externalities of the past. According to biblical law, negative externalities may last four or five generations because each succeeding generation gets a chance to rectify the sins of the parents through overcompensation (cf. Exodus 20:5). The concept of divine grace as expressed in Exodus 20:5 is essentially grace with regard to the children of a generation of sinners. They are innocent when they are born, just as are the children of the generation of victims. The children of the sinners should not be punished for the collective guilt of their fathers (c.f. Ezekiel 18:1-3 where the proverb, "The fathers eat sour grapes, and their childrens' teeth are set on edge" is utterly rejected by the Lord God). But these children have to rectify the situation created by their parents. Collective guilt thus manifests itself in the failure of a generation to rectify the failures of past generations.

Figure 6–4
The relations of Christian citizenship in a geographical framework

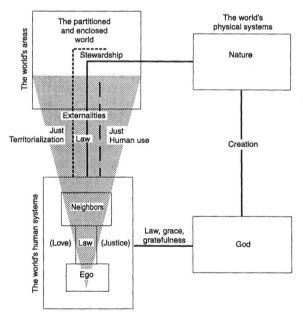

Divine grace has implications for geographic research and education by those with a Christian background, especially with regard to externalities. Members of each generation must be aware that they are fellow citizens (in-laws) and, at the same time, aliens in the eyes of citizens on the other side of the border. This is the Janus head of biblical citizenship. As citizens, they ought to know their own history and geography because they have to pass judgment on the geo-historical situation in order to determine whether they inherited a righteous regional differentiation, and if not, to rectify past failures with regard to aliens on this side of the border as well as to aliens living on the other side. As aliens, they have to show patience in order to give citizens a chance to compensate for negative externalities. Hence, Christian geographers have to interpret and judge the regional inequality from the perspective of thick morality and the biblical form of universal morality. As teachers, they have to teach (prospective) citizens to interpret and judge human decision making with regard to economic land use and spatial organization, resulting in a partitioning of world areas in relation to socioeconomic and demographic processes (Figures 6–2 and 6–4). A Christian worldview is basically a retrospective one, and Christian scholarship is an explanatory and interpretive activity concerned with past and present patterns. But prediction is also relevant to this science since good citizens have to understand the effects of their decisions. Citizens are good stewards if they create a regional differentiation that is righteous. In that case, the aliens of this generation would have received their due and the citizens of the next generation would not have to compensate aliens of the next generation for the failures of their generation. Under those conditions, we would come very close to the kingdom of God on earth.

Aliens in a Foreign Land

A Christian worldview as prerequisite to righteous citizenship requires a critical distance from the present world (John 17:16; Rom. 12:2; 2 Cor. 6:11-18). Crital distance is one of the essentials of Christian ethics. It denotes the mental distance one must have to be an independent judge of conduct and events (Rich 1980, 22). It is the basis of Christian freedom to act in obedience to the will of God (Wendland 1971, 42). In view of the critical distance that a Christian citizen must maintain with regard to the recent past and present, it is unlikely that this view of citizenship forms an acceptable educational target in any society other than a democracy. But even in a democracy, critical distance is considered taboo, at least if we correctly interpret Heater's exposition of a citizen's involvement as depending on its context. He formulated the educational policies of secular citizenship in a participatory-democratic society as follows: "development of an individual child's attitudes *and* social homogenisation, and knowledge and understanding of the system and participative skills" (Heater 1990, 213). In a Christian worldview, social homogenization is an undesirable educational aim if it stimulates the cultural homogenization of a basically plural society. Knowledge of the biblical variant of thin morality, and its confrontation with present thick moralities, is a better way to stimulate good citizenship in a plural society. Knowledge and understanding of the system and participative skills are necessary but not enough to become a good citizen in a biblical sense. One must have knowledge of past, present, and future patterns of regional differentiation and the negative externalities in these patterns. Therefore, regional and urban geographical research and education that is inspired by the Bible must incorporate the antithetic principles of universal law (equality) and biblical law (inequality) and demonstrate how they operate in space.

The ethical judgments carried by a Christian worldview are easy to integrate into the conceptual structures of geography. The key concept of externalities is present in most theories, albeit often designated by other terms and used as an explanatory category and not meant to serve as an instrument of moral judgement. However, the opportunities to arrive at judgments will require a change in priorities in geographic research and the school curriculum. The emphasis on economic geography (and location studies) in a world that aims at global competion should be abandoned (cf. the educational aims formulated by the Geography Education Standard Project (National Geographic Research and Exploration, 1994). Geography should be focussed on the central concepts of citizenship and externalities. Citizenship requires a strong emphasis on historical, political, urban, and regional geography. An education informed by a Christian worldview adds elements or aspects to the accepted standards of geographic education that promote the formulation of ethical judgments about the operation of human systems. Historical geography, for example, should not only lead to a better understanding of human environments in the past; it should also question whether the past was wrong or right (a colonial past, the creation of centers and peripheries) and if and how such faults have been compensated. Urban and regional geography's models of types of regions and their formative processes (National Geographic Research and Exploration 1994, 72) should be selected by the ways in which externalitics are included. In order to prepare future citizens for participation, political and cultural geography should also expand on such externalities by elucidating "the forces of cooperation and conflict" (National Geographic Research and Exploration 1994, 210, Standard 13) and by evaluating them from the viewpoint of justice.

The concept of externalities requires that thematic geographical theories and models should be applied in geo-historical and geo-political contexts on the local, regional, national, and international levels. The students of today are the citizens of tomorrow. They may become planners or politicians. But whether planners, politicians, or the general electorate, everyone has to learn about externalities. In a democracy citizens have a choice whether to accept *compensation* instead of *competition* as the guiding principle of planning policies. If they accept compensation, the second choice is to apply the French revolutionary variant of equality or the biblical variant of inequality. But as long as knowledge and skills to compete in a global economy are essential parts of national education goals, Christian geography teachers will be "aliens in a foreign land."

Notes

[1] The author is grateful to Mrs. Nancy Smyth-Van Weesep and Prof. Dr. Henk Aay for correcting his English, to Prof. Dr. S. Griffioen and Dr. Hoekveld-Meijer for their comments on earlier versions of this chapter, and to Prof. Dr. H. Tieleman for his advice during the preliminary phase of my exploration of this unfamiliar field.

Chapter 7

Christian Worldview and Geography: Christian Schools in the Netherlands 1900-1960

Henk Aay

> If geography is to be Christian then this can only come to expression
> in the consideration of the subject material. . . . Does it have any
> influence on the results if one describes France as a Christian or as
> an atheist? Does our world and life view come into the discussion
> when we prepare a monograph on Argentina? Can people treat the
> city of Rotterdam in a Christian and in a non-Christian manner . . .?
> How would one Christianly deal with monsoons, glaciers, and ocean
> currents? This is equally as difficult as putting together a Christian
> chemistry or a Christian physics! (Vermooten and Sleumer 1929-30,
> 262-63).

Introduction

IN THE WORDS of Albert Wolters (1985a, 2) a worldview is "the comprehensive
framework of one's basic beliefs about things." Every individual and cultural
movement possesses a worldview and every institution incorporates one. That
worldviews shape and direct geographic perception and behavior is a well-
accepted tenet within geography today. Such worldviews funneled through
national and cultural traditions and pioneer scholars also mold the intellectual
constitution of geography as a whole; we regularly speak of distinctive national
traditions and schools of thought and practice within geography. Geographic
education, too, is constructed from within the worldview of a culture or a
subculture. This study examines the impact of a Christian worldview on
geography in education in the Netherlands during the first sixty years of this
century.

Confessionally directed education, both elementary and secondary, has been
a significant part of the Dutch school system since the late nineteenth century
(Gilhuis 1974; Janse 1935; Langedijk 1953; Messelink 1994; Stilma 1987, 1995;
Van Brummelen 1986). Since 1920, Protestant and Catholic schools have
received the same level of state funding as the public (neutral) schools, thus
establishing a religiously pluralistic public school system. These special schools,
as they were called by the government, have always made up a significant share
of national enrollment. In 1930, for example, special elementary schools
enrolled more than 700,000, some 62 percent of all students; in the same year,
nearly 27,000 students were enrolled in special secondary schools, nearly 45

percent of all students. Special schools were either Christian (the name for Protestant schools) or Roman Catholic; each of these confessional orientations has constituted a broad enrollment stream within the national school system, with the Roman Catholic schools consistently leading in the enrollment count. In 1969, for example, when Roman Catholic and Protestant school enrollments were reported separately in the Dutch census, 27.5 percent of all elementary school students attended Protestant schools and 43.5 percent attended Roman Catholic Schools. Broadly considered, the Dutch educational system is a three-cornered system with equal weight given to each corner (Centraal Bureau voor de Statistiek 1940, 1970).

This chapter examines one of these three corners: the Christian schools. As mentioned above, this is the (inapt) name given to Protestant education. In the Netherlands, Protestantism was decisively shaped by John Calvin (1509-1564), the French reformer, and his followers. The Christian schools, however, are more accurately described as Reformed Christian schools or as Calvinistic schools. To understand the nature of these confessionally directed schools in general and their geographic education in particular, it is necessary to review some salient features of this neo-Calvinist worldview and culture of the late nineteenth and early twentieth centuries.

The revival of Dutch Calvinism in the second half of the nineteenth century took place on a broad cultural front. It was no longer restricted to Reformed theology, faith, and the church but expanded to a Calvinistic world and life view decisive for every arena in society. Its principals were not only clergy but also educators, labor organizers, journalists, lawmakers, and academicians. Under the vigorous and charismatic leadership of Abraham Kuyper (1837-1920), journalist, statesman, and theologian, who also served as prime minister of the Netherlands, Dutch neo-Calvinists established parallel Christian organizations in every sphere of society. Included were Christian schools and their national organizations, a political party, a Christian university (the Free University), labor union, and press. Not only did this confessional pillar structure daily life for Reformed people in education, politics, labor, church, and the press, it also reached in to organize more commonplace activities. If someone enjoyed traveling, there was the Dutch Christian travel organization; if someone was interested in learning Esperanto, there was the Dutch Society for Christian Esperantists (Woudenberg 1929-30)! Neo-Calvinism encouraged Reformed Christians to form separate organizations for the many and diverse areas of human endeavor.

National law and policy determined the subjects to be included in the curriculum by grade level and by number of teaching hours per week. Already in 1857 for primary education and in 1863 for secondary, geography became a compulsory subject in Dutch education. Grades one through three already included some preparatory geographic education related to basic concepts as well as to hometown and surroundings. The remainder of the elementary school years included regular weekly geography classes, completing a cycle that focused on the Netherlands (including its overseas areas), Europe, and the world. The first three years of high school normally repeated this general geographic survey of the entire globe with the same regional emphasis, reflecting the Netherlands' place in the world but with greater intensity (six to eight class periods per week). Depending on the type of high school, the two or three senior years gave opportunity for more specialized topics in geography such as physical, social, and economic geography as well as regional studies. In general, these were elective courses meeting two to four class periods a week.

With geography compulsory throughout much of grade school and high school, the Netherlands made a large national investment in geographic education. This reflected the economic expansion and cultural reawakening of the Netherlands beginning in the latter part of the nineteenth century; both within Europe and the world, a tiny Netherlands actively sought expanded external relations. A country much dependent on others for its well-being required a geographically educated populace. The national importance of geography was reflected in a large community of state-certified geography educators, a growing and diverse body of geography textbooks and other curriculum materials, and the growth of teacher-training colleges and university-level geographic scholarship and education. As with all social institutions during this period, the affairs of geographic education in elementary and high schools, teachers colleges, publishers, and university geography departments were structured by the confessional pillars of Dutch society. Roughly one third of Dutch grade and high school students received their geographic education within the broad contours of a Protestant Christian worldview.

Neo-Calvinism and Geography

Geographical Dimensions of the Worldview
Neo-Calvinism as a Christian worldview manifested a number of distinctive features (Wolters 1985b, 1995). I will restrict myself to those aspects of neo-Calvinism that are of significance to the field of geography. One distinctive feature of a neo-Calvinist worldview may be summed up in the phrase: Life is religion. Not reserved for the life of faith and spirituality, religion is the central power and direction that always and everywhere encompasses a person's entire life—thoughts, feelings, and actions; it extends its influence over all of humanity, over family, society, scholarship, and state. Moreover, religion extends beyond formal, organized, explicit religions such as Islam and Christianity to include comprehensive implicit value systems such as Marxism and materialism. For geography, the centrality of religion meant that the religious life of a country or region should not be neglected in a geography text or lesson. And, more to the point, the religion of an area should serve as a keystone to understanding its geography. Indeed, a common complaint in reviews of geographic literature in magazines for Christian education is that the author says nothing about the religious life of the people and places described.

Another distinctive feature of neo-Calvinism is the idea of the *antithesis*. This refers to the fundamental division and opposition between belief and unbelief. Culturally, this meant that Reformed Christians everywhere were required to counteract secularism. This led to the establishment of Christian schools in which geography also would be taught—thinking Christianly about geography, journals, and magazines for Christian education; writing geography curricula and textbooks for such schools; and interacting with other geographic educators within Christian teachers' organizations. It is this record that constitutes the primary resource for this research. A significant number of journals and magazines for Christian education and educators were available especially in the pre-WWII years. In these publications, matters relating to geography and geographic education were included: book and textbook reviews as well as articles on methodology, curriculum, pedagogy, and educational policy. In addition, there are a number of geography textbooks, readers, and other classroom materials written expressly for Christian education.

It is well-known that Calvinism stresses God's sovereign rule by which His creation is governed. This theme is further elaborated in neo-Calvinism under the idea of the law order. This is a complex notion of a constant cosmic order; constant lawlike structures hold for all creatures whether these be hurricanes, butterflies, human thoughts, or commodity markets. In the human sphere such principles take on the character of norms. Both the Scriptures and creation reveal these normative principles. For neo-Calvinist geographers the significance of the creation order was felt in several ways. Following Karl Ritter, it gave opportunity to relate the physical geography of the Earth to its suitability and meaning for culture (configuration of continents and oceans, distribution of land and sea, the dynamic balance of the Earth's crust, and so forth). All this pointed to God's wise design and His order for the cosmos. More generally, geographic education would instill deep feelings of wonder, awe, and love for the creation and, more importantly, for the Creator who fashioned and upholds everything with power and wisdom (Livingstone, above, chap. 1).

Neo-Calvinism also consistently expressed an appreciation for cultural development through the principle of the *cultural mandate*. Humankind is commanded by God to develop the potential within the creation. The products of such cultural development whether new forms of transportation, consumer goods, or social institutions are part of the goodness of creation that no amount of perverse and distorted use by people can remove. For neo-Calvinist geography educators, a commitment to the cultural mandate meant little patience with any form of environmental determinism or social Darwinism. Among those committed to a human ecological definition of the field (most were not), possibilism (a cultural-historical methodology) rather than environmentalism pointed the way. Moreover, the development of the creation by people is a cultural requirement that produces a structural differentiation of societies.

A last feature of this Christian worldview with relevance for geography can be summed up by the well-known phrase of Herman Bavinck (1826-1909), the most important theologian of Dutch neo-Calvinism: Grace restores creation. This means that the Christian gospel speaks of a renewal and a reestablishment of created reality according to the original design of the creator. The word creation in this phrase is not restricted to what is commonly referred to as the natural world. Rather, it also takes in all of human culture. *Grace restores creation* makes Christianity a "cosmic salvage operation" to use Al Wolters' apt phrase (Wolters 1995, 44). In every place where sin and evil have darkened the original goodness of the creation, for example, through hunger, disease, pollution, and idolatry, Christians must strive to restore the original goodness of the creation.

This worldview theme was also applied to geographic education under the heading of missions. While the topic of missions was also included in required courses in religion, it is significant that materials related to world missions were also taught as part of the geography curriculum in the Christian schools. Such mission activities included not only evangelism but also institutions of mercy, healing, and education. Even further, they included the changed economic and social circumstances that the mission fields fostered. As a result, the geography of a country or continent as taught in Christian schools included significant mission places alongside other denominations of significant places.

Scriptural Principles for Geography
The above features of neo-Calvinism as a comprehensive worldview relevant for geography were historically developed on the basis of the Scriptures. Some neo-Calvinist scholars also sought more direct connections between the Bible and specific areas of human endeavor by identifying apposite scriptural principles. These would direct the interpretation, evaluation, and understanding of findings, events, and experiences. As members and leaders of a subculture, geography educators would already represent and apply many such principles unselfconsciously; these were part of the air they breathed. Nevertheless, those who concerned themselves with the foundations of geography sought to make explicit and systematize those scriptural principles intensely relevant for the field. Once such principles were appropriated by students and scholars alike, they would steer the understanding and interpretation of the geography textbook, curriculum, or research. A. Van Deursen, arguably the most influential and scholarly neo-Calvinist geographer of this period, more than any other person, focussed on setting forth scriptural principles for geography. In three publications that span his career and that discuss the foundations of geographic education, he makes these principles a centerpiece of his work (Van Deursen 1921, 1937, 1951).

I will only cite several examples of such scriptural principles for geography in order to portray their character; some (missions) have been previously adumbrated; several others will be elaborated on in succeeding sections. Some principles are very broad and seemingly without much significance for geography per se. For example, Van Deursen (1921, 21) holds to the principle that in a religious and ethical sense the earth is the center of the universe in spite of its physical insignificance, and that, accordingly, the earth is the only place for creatures such as human beings and the only venue in the universe for the struggle between good and evil and the establishment of the kingdom of God. Many other principles stem from a decidedly literalist hermeneutic, divide general from special revelation, and are therefore much more restrictive for geography. Van Deursen listed as one of the scriptural principles that humanity's fall into sin profoundly altered the physical earth and that volcanic craters and polar regions serve as evidence (Van Deursen 1921, 21, 28). It appears that the general and the comprehensive identifying features of neo-Calvinism were far more positive than the more specialized scriptural principles for developing a Christian orientation in geography.

Sociography: The Reigning Paradigm for Geographic Education in Christian Schools

> The scriptures teach us, do they not, that humankind is the crown of the creation, who mirrors the image of God even in his physical body. From this it follows, that geography which limits itself to ground and water, to climate, flora and fauna misses its coping stone, that this field of study is for us, . . . nothing less than geography and ethnology [land- en *Volken*kunde] while the latter is the main issue. . . . When the other countries of Europe are discussed, here, too, human life has to constantly form the keystone and in the center of that existence there stands the reality of religion (Wielemaker, 1915, 268, 275).

The foundations of geography and geographic education in the Netherlands during the early twentieth century were anything but certain. Within academic geography the methodological currents swirled around Ratzel's environmentalism, Hettner's chorology (areal differentiation), Sauer's (and his influential Dutch student Jan Broek's) landscape approach, Vidal de la Blache's society-milieu, and Steinmetz's sociography. Geographic education at the primary and secondary level, much affected by the crosscurrents in academic geography, was seriously looking for alternatives to the educationally unproductive "bays and capitals" rote place-name and place-description learning. At this time, the two principal centers of geographic scholarship and education within the Netherlands were at the public Universities of Amsterdam and Utrecht. Each developed a distinctive school of thought and strong loyalties. In Amsterdam, S. R. Steinmetz developed an orientation that was known as sociography. It attracted, among others, geography teachers interested in advanced degrees. More significantly for our purposes, the sociographic orientation at the University of Amsterdam had a major influence on leading geographic educators at Christian schools; important figures such as W. H. Vermooten, W. Sleumer, and A. Van Deursen graduated under Steinmetz with Ph.D.s. Equally significant, the first geography textbooks incorporating a sociographic approach were written by Christian geographers.

What constituted a sociographic orientation to geography and why were geography teachers in general, and teachers at Christian schools in particular, attracted to it? Unlike many other philosophical orientations within geography that had international scope, sociography was a more limited academic movement without much broader influence. It proved to be a blind methodological alley in the history of geography. Its impact was greatest in Dutch geographic education. The easiest way to understand the meaning of sociography is to replace the term with its present-day operational synonyms: ethnography, cultural anthropology, and social studies. Sociography was the (comparative) study of societies and cultures. At a time when the individual social sciences were still largely absent from university faculties, geography via sociography provided an umbrella for comparative ethnography. Geographers with a sociographic disposition wished to include purely social and cultural phenomena in geographic scholarship and education. It is important to remember that until this time there was very little human-cultural content in the geography curricula of elementary and secondary schools; place names and descriptive relief dominated the geography lesson. Steinmetz and his followers insisted on admitting population, livelihood, law, race, social institutions, religion, and so forth to geography, which would examine these social phenomena in their interrelationships in a particular culture. Geography educators welcomed this new orientation. It met their dissatisfaction with the customary curriculum and, at the same time, preserved a synthetic and unified approach important for geography as a school subject.

As the decade-long debate within the pages of the *Tijdschrift voor het Onderwijs in Aardrijkskunde* [*Journal for Geographic Education*] attests, there was deep division and sharp disagreement about sociography (Fahrenfort 1926; Kuperus 1926; Zondervan 1926; Hanrath 1926; Steinmetz 1930; Cohen 1931a; 1931b; Broek 1933; Kruijt 1934; Kuperus 1934; Barten 1936). The disagreement was between the Amsterdam and the Utrecht schools, between geography educators and geography professionals. The Utrecht school followed the French geographic tradition of Paul Vidal de la Blache, which focussed on the relationships between society and milieu. In the main, Christian geography educators were more attracted to the Amsterdam school of sociography than to the Utrecht school. By its detractors, sociography was called a school subject and not a scientific discipline. But the Achilles heel of sociography was that it did not bring a geographic approach to the study of social and cultural phenomena; sociography did not usher in a social (human) geography. Sociography left geography out. The products of sociographic investigation were no different from descriptive and comparative sociology, ethnography, political science, or economics. This can especially be seen in the Ph.D. dissertations written under Steinmetz' supervision and in geography textbooks written from this point of view.

Van Deursen, completed a dissertation in 1931 under Steinmetz on the topic of the Messiah among North American Indian cultures. Van Deursen certainly knew the traditional geographic points of view (nature-society, the human landscape); he had taught geography at the high school level for more than a decade and was widely read. His dissertation, however, is a straightforward, ethnographic study without any geographic analysis. Romkes (1924a, 1924b, 1927), another Christian geographer and high school teacher, wrote the first textbook series that featured human and cultural materials entitled, *Beknopt*

Leerboek der Aardrijkskunde [*Concise Textbook for Geography*]. Included in the treatment of countries in each region are extensive sections entitled "people" and "society." They contain descriptions of art, religion, language, economic development, government, and industries among others. Yet, virtually none of these materials show geographic distinctions. It may be argued that if Ph.D.-level research could not or did not see the need to interpret these phenomena from a geographic point of view, then geography textbooks could hardly be expected to accomplish this. Indeed, present-day world, regional-geography school textbooks often fall short of the mark in this respect also. The sections of these geography texts that cover people and society read like social-studies school books.

Neo-Calvinist geography educators thus played an important role in bringing human states of affairs into the center of geography. In this they led a larger movement in geographic education which, by and large, initiated and welcomed such changes. While others based the wisdom of such curricular change on the relative importance of human verses natural phenomena in the world ("we don't care about the height of mountains but we do care about the Negro problem" [Romkes 1924a, IV]).[2] Christian geographers had their own more nuanced reasons. These included placing the spiritual driving forces of a society onto the center of the stage and recognizing that humanity is the primary agent in advancing such cultural ideals.

At the same time, by embracing sociography, Christian geographers, together with the larger community of geography educators, isolated themselves from academic geography and headed down a methodological cul-de-sac that quickly came to an end in professional geographic scholarship. The common and telling geography textbook title, *Land en Volkenkunde* [the study of land and people] did not refer to the interplay between people and land but to two more or less separate compartments. The land part of the title referred to geography (physical landscape and land use) and the *volk* part referred to ethnography. By insisting on the inclusion of diverse societal phenomena without relating these to land or to geographic frameworks, these educators left geography for the school subject of social studies.

Roads Not Taken

The priorities that Christian geography educators brought to the subject could also have been applied to other paradigms within geography, notably to the central themes of landscape, chorology, and society-milieu. A small beginning can be seen in the work of G. De Jong, the first professional Ph.D. geographer appointed at the neo-Calvinist Free University of Amsterdam during the late 1940s, nearly seventy years after its founding. De Jong (1955; 1962) accommodated his neo-Calvinism to the chorological concept of Hettner and Hartshorne in his well-known work (translated into English), *Chorological Differentiation as the Fundamental Principle of Geography*. He writes (in his typical dense philosophical style):

Divine powers carry the distinguished unities of vertical and horizontal totality in each region and in the mutual spatial coherence within the entire Earth's surface and give them direction. And when in every totality within each region and among the regions in the mutual coherence of the entire Earth surface the shadow of the curse can clearly be seen, the richness of the creation is not destroyed by this; on the contrary, in the various spatial horizontal and vertical differentiations the creation by God's grace comes to development (De Jong quoted in Van Deursen 1955-56, 572).

De Jong's synthesis of neo-Calvinism and neo-Kantian chorology remained largely undeveloped, with no followers, and came at the end of the period during which Christian approaches to geography and geographic education were under consideration. It is ironic that just at the time when the Free University was (finally) establishing geography as a program of study and department in the early 1960s, questions relating to Christian worldview and science began to wane at the University and in Reformed circles. The products of geographic scholarship from members of the geography faculty (a Ph.D.-granting department) at the Free University from the early 1960s to its closure in the mid 1980s by and large do not bear the imprint of a neo-Calvinist worldview nor of an explicit Christian worldview in general. The orientation of the first faculty members was in the direction of the Utrecht school and the sociographic paradigm was not represented. Moreover, the worldview orientation of key founders of the department, men such as M. W. Heslinga and G. A. Hoekveld, was not neo-Calvinist but more broadly Reformed.[3] It is interesting to speculate what the level of development of Christian thinking about geography and geographic education would have been had there been a neo-Calvinist Steinmetz-type figure and a geography program at the Vrije Universiteit during the 1920s, 1930s, and 1940s.

What cannot be denied is that geography teachers at Christian schools could not look to their university for direct leadership and philosophical direction. As a result, their own leaders, many of whom had taken advanced degrees, took the task of reforming geography upon themselves. Because they were teachers, the accomplishment of this task remained bound to the needs of geographic education more than to the foundations of geography. They turned to Steinmetz and sociography, not because the former was a Christian scholar and the latter was developed on the basis of a Christian worldview but because this paradigm could accommodate their priorities and because it was relevant for the curriculum and the classroom. It must be said that translating academic paradigms such as human landscape and chorology into an integrated school geography curriculum would have been a daunting task. This points to the enduring challenge and difficulty of traffic between academic geography and the needs of geographic education. Geography teachers in Christian schools discovered in sociography, a very broad program that would more readily transfer to the world of the classroom.

Another road not taken to a Christian orientation in geography, one that can only be briefly mentioned here, is the neo-Calvinist philosophy of Herman Dooyeweerd and Dirk Vollenhoven developed during the 1920s and 1930s, concurrent with the conceptual stirrings in geography. This encyclopedic Christian philosophy constructed a comprehensive and integral ontology and epistemology that was very serviceable for any field of learning. While a number of disciplines benefited from a working relationship with this Christian philosophy and with neo-Calvinism in general, geography was not one of them (Van Os and Wieringa 1980).[4] In all the writings about geography within the Christian school movement there is scarcely a single reference to this system of Christian philosophy. Here was a missed opportunity.

Worldview and Paradigm in Motion: Contours of Neo-Calvinist Geographic Education

Geography Textbooks for Christian Schools
If geography shaped by a Christian worldview was to be effective in the classroom, then the teacher, the curriculum, and the textbook were on the front line. Normally, teachers in Christian schools were trained in Christian teacher colleges and certified by the state. The broad curriculum was set by the state. Individual schools and teachers decided what textbooks and ancillary materials would be used by students. In some subjects, such as religion, reading, literature, and history, textbooks written for Christian schools were commonly used. Such textbooks were not required, however. The same textbooks as those of the public schools were used, with the teacher expected to provide the needed Christian perspective. In other (sometimes called "neutral") subjects, such as the natural sciences and mathematics, such Christian textbooks generally were not deemed to be as necessary because these were more indifferent to a Christian worldview. Geography fell between these two extremes, partly because school geography was just emerging from rote place-name learning and descriptive topography, and partly because the field included both natural and social-science content.

Not many geography textbooks were written specifically for the Christian schools, nor were many textbooks written for schools in general from a Christian worldview.[5] Most were published in a brief period during the later 1920s, shortly after the Christian schools received state funding. They were an untested innovation that proved remarkably short lived, saddled with small markets, poor receptions in educational circles, and weak commitment to the need for such materials. They were produced in a variety of formats: conventional textbooks that systematically organized materials, such as Romkes' (1924a, 1924b, 1927) *Concise Textbook for Geography* and Van Deursen, Overweel, and De Vries (1926-1939), *The World and Those Who Live in It*; geography readers that told stories (often in first person plural) to make geographical knowledge easily understood, such as Hoogwerf and Baarslag's (1929a, 1929b) *Geography Reader for Christian Education* and Lankamp's (n.d.) *Geography in the Christian School*; and, finally, geography anthologies that by region assembled

items published elsewhere, for example, Wansink's (1928) *Geographical Anthology: Of Lands and People (For the Christian Schools).*
Hoogwerf's and Baarslag's (1929a, 1929b) *Geography Reader for Christian Education* was enthusiastically received by Christian education reviewers who hailed this eight-volume set (plus three map booklets) as the first complete and original geography curriculum for Christian elementary education. Like most such textbook sets, this reader was organized by world regions with a predictable emphasis on Europe and the Netherlands. Methodologically, this series took its inspiration from Jan Ligthart, a pioneering and influential public school educator and writer who enlivened textbooks with stories and imagination and who stressed the moral upbringing of students (Cremer and Ligthart 1909); Hoogwerf's and Baarslag's geography reader was Christian education's answer to Ligthart. The students travel through and experience the geographical features of places and regions in these stories.

Throughout the series, the Christian orientation is explicit and accented and is focussed on providence and piety. Earlier, Hoogwerf (1918) provided a blueprint for these textbooks in which these two qualities are more systematically presented. Physical geographical features consistently disclose God's design and purpose (for humanity) and, likewise, human landscape features are part of God's creation and grace; here are educational objectives for a lesson on rivers (the Rhine as model):

> We get up on the dike with our young people. They should regard the river as a rich gift of the Creator. That stream brings fresh water from God's treasury. Far away in the mountains He placed His glaciers, by means of which He provides a perfect distribution in all seasons. Moreover, He gave the rich mud year after year and demonstrated that He labored for us for centuries. And that still was not enough. That stream took on the task of a mighty broom, which provided a daily cleaning of much of the rubbish of humanity, which in time would even spoil the oceans, had she not received preserving salt in her lap at earth's dawn and continuing still. Add to that the potential of the river for commerce and shipping and her riches for the fisherman and—in this geographical way our boys have discovered God's footsteps (Hoogwerf 1918, 215 16).

In the built landscape, the students are routed through places of Christian significance (churches, institutions of Christian mercy), and even ordinary landscape elements, such as bridges, while testifying to human technical power, speak to the potentials that God has structured into His creation.

Each of Wansink's (1928) six-volume geographical anthology contains more than thirty short selections by different authors. Many are ethnographic exploration accounts cast in a rather romantic and heroic mold. The Christian orientation is far more muted than in Hoogwerf's and Baarslag's (1929a, 1929b) *Geography Reader.* It consists of scattered attention throughout many of the pieces to the work of Christian missions and the presence of Christianity in the region under discussion. Equally noteworthy, but not surprising, is the attention paid to the life of religion in these selections when compared to other dimensions of culture.

Romkes' (1924a, 1924b, 1927) three-part *Concise Textbook for Geography* was not expressly written for Christian schools although they were part of its intended market, and a Christian worldview is both implicit and explicit in the work. In his treatment of Palestine, in his emphasis and discussion of religion, in his opposition to evolutionism, and in his evaluation of social conditions, Romkes explicitly discloses a Christian orientation (Cohen 1924, 34). More implicit and more formative is the text's sociographic paradigm with its own Christian underpinnings as noted above. At bottom, Romkes (1929-30, 423) wanted to combat the materialism that was also appearing in human geography:

> In connection with the Congo I cannot be satisfied with naming the imports or exports. Next to the tropical rainfall, I *must* point to the dreadful social conditions to the spiritual needs of the black population and what is being done for their improvement, alleviation and development.

Although the series was recognized as charting a new direction in geographic education, it was roundly censured for numerous inaccuracies, imbalanced regional coverage, and educationally inappropriate text. It failed in the Christian and public-educational marketplaces (Cohen 1924).

The publication of the five-volume series *The World and Those Who Live in It* (taken from Psalm 24:1) by A. Van Deursen et al. (1926-39) gave occasion for the most intense and lengthy debate about Christianity and geography and showed that there was deep division within the community of geography teachers in Christian education. The divergence of views was due in part to the differences in outlook between Christian educators who were neo-Calvinist (Kuyperian) and those who were members of the Reformed church and who took a more theological-ecclesiastical line. The argumentation lasted three years (1929-31) and took up more than one hundred pages of the journal, *Christian Secondary Education.* After this lengthy, personal, and heated debate, there were no more geography textbooks published for the Christian schools.

Christian Perspectives on Geographic Subjects
The central charge leveled against this series of geography textbooks, and, by extension, against all Christian geography textbooks, was surprising and uncharacteristically un-Calvinistic. W. H. Vermooten and W. Sleumer, geography teachers at prominent Christian high schools and, like Van Deursen, followers of Steinmetz, indicted these textbooks for violating the separation between science and religion in their efforts to construct a Christian geography and a geography curriculum for Christian schools. To each of the questions in the opening quotation of this paper about the difference that a Christian approach might make, Vermooten and Sleumer answered a resolute no. Operating from

a positivist worldview, which separated facts from values, these geography teachers were committed to a value-free disciplinary knowledge driven by an objective scientific method and a critical experience epistemology. The content of the subject matter of geography would be the same regardless of worldview, religion, philosophy, or any other metaphysical condition. That is not to say that Vermooten and Sleumer did not support the case for Christian schools or considered realms of values unimportant or even unrelated to geography; it is only that such things, in their view, do not belong to the school subject of geography but in a Christian school to the subjects of religion, Bible, and church history. Vermooten and Sleumer were both positivists and Christians, a not uncommon worldview alliance in scholarship and education.[6]

For Vermooten and Sleumer, these geography textbooks themselves were proof that a Christian worldview could not author a Christian geography. Stripped of their illustrative Bible texts, confessional prolegomena, and descriptions of Palestine, *The World and Those Who Live in It* volumes were essentially the same textbooks as those of the public schools. Repeatedly, these critics, who were not popular in the circle of Christian educators, rightfully asserted that for there to be a truly Christian geography, the approach must produce a distinctive treatment of geography's subject material itself: "It should be obvious to everyone that Christian subject-matter knowledge only makes sense when it is a question of *another focus* with respect to the treatment of the assigned subjects"(Vermooten and Sleumer 1929-30, 262). Not introductory scriptural principles, bible references, and Christian institutions but especially the nitty-gritty physical and cultural phenomena under study would need to be seen in a different light. Such Christian geographic knowledge of France, Rotterdam, Argentina, glaciers, monsoons, and ocean currents (the list of topics in the opening quotation) Vermooten and Sleumer could not find in this series of textbooks and, indeed, it is not present. This failing is not surprising to them; rather, *The World and those who Live in It* is proof that such a project is fundamentally flawed. From their positivist outlook there is only one geography, empirically unassailable by metaphysics. While biblical starting points and religious confessions may lead one to see this neutral geographic knowledge in a different light, these do not belong to geography proper.

Vermooten and Sleumer challenged and dared their critics to produce a geography of a place or phenomenon written from a Christian worldview (1929-30, 408). Such studies were then as now not available. This is not because of the ontic impossibility of such scholarship as a positivist would assert but, rather, because then as now there was/is no Christian geographic mind, and no scholarly institutions to cultivate and advance such a mind. Without a Christian philosophy of geography, without examples of Christian scholarly geographic research, without academic leadership from Christian universities, and last, with a geographic methodology freighted with descriptive inventory, it is not at all surprising that educators could not write synoptic geographic textbooks that got beyond general confessional principles. Although Van Deursen, Overweel, and DeVries (1926-39, vols. 2 and 4, Preface) were convinced that a Christian worldview would affect the treatment of geography's subject material, the Dutch neo-Calvinist academic and educational community did not possess the resources necessary for such an inner reformation of the field of geography.

Christian Subject Material for Geography
Another, arguably less ambitious but much more common approach for bringing a Christian worldview into educational materials for geography seizes on the selection standards for materials. For the authors of *The World and Those Who Live in It*, the choice of subject materials was another significant nexus between Christianity and geography but Vermooten and Sleumer dismiss this as a trademark and insist that if geography is going to be called Christian it can only do that by its treatment of all its diverse subject material:

> However they [the authors] go further and they believe that also the choice of subjects in Christian geography has to be different than in geography in general. We doubt whether this is correct. After all, whether a branch of knowledge is Christian or not the territory that it cultivates remains the same. Indeed, it may not make a claim to the name science, if it knowingly disregards certain parts. Possibly, the authors meant that they pay attention to missions and religion more than others, but then we have to point out to them that in this they do not in any way produce something that is fundamentally different. Christian geography has to be something else than geography in general plus something else. An agnostic author can do this just as much from an interest in the religious life of peoples and in our day that is not at all uncommon (Vermooten and Sleumer 1929-30, 262).

A Christian approach to geography and geographic education that highlights or escapes into a certain kind of subject matter remains prevalent today. Christian geographers who do not limit their faith to the private domain and worship (many do); those who embrace a working relationship between their Christian faith and their geographic scholarship sometimes find their refuges in the geography of religion, the geography of the Bible, and even in the philosophy of geography. As Christians, they are more likely to study the journey to worship or the location of shrines than the journey to work or the location of crops.

Although they themselves deny the thesis, Vermooten and Sleumer provide a classic neo-Calvinist position on the relationship between Christianity and geography: The entire canvas of geography should be illumined by a Christian mind, not only charitable organizations and religious buildings but also, with equal significance, soils, solid waste, and subsistence agriculture. Exactly how this was to be accomplished remained unclear, but this principle stood as a central feature of the blueprint for constructing a Christian geography. While a secular scholar might also wish to include the religious practices and institutions of a region, such phenomena would be cast into an entirely different light by a Christian geographer (Van Deursen and De Vries 1929-30, 415). As positivists, Vermooten and Sleumer insist that even this subject category would not result in a distinctive treatment by Christian geographers if the scientific method were employed.

While Vermooten and Sleumer are correct when they point out that a Christian approach to geography cannot be a church and missions icing on the cake of a neutral geography, what about their dismissal of any selection criteria for subject material in geographic education and scholarship? Is geographic education steered by a Christian worldview complete when a Christian mind is engaged in textbooks, teaching, learning, lesson plans, and curriculum across the entire common array of subject material? Or, does Christian education at all levels have a no less important primary task of telling and situating the students in the story of Christianity? In the words of Reformed theologian John Bolt (1993, 221): "The Christian School is a place where the civilizational significance of the Christian religion is narrated, studied and practiced." Christian scholarship and, especially, Christian education, has a particular responsibility to study Christianity in all its cultural manifestations and relationships. This helps build Christian cultural literacy which is knowledgeable about "the grand narrative of God's redemptive mission" (192) and committed to active citizenship in God's kingdom.

In Christian education, such an orientation by its choice of subjects results in quite a different curriculum from that of the public schools even if their programs of study include the study of religion. Most obviously, subjects such as Bible and church history make their appearance, but the same selection standards are also applied in history, geography, and literature. In Christian scholarship in general, such selection standards are not nearly as forceful, but it must be emphasized that here, too, a Christian academy or scholar has a special responsibility and privileged position to study Christianity in all its dimensions.

The distinctive cargo of geographic education in Christian schools in the Netherlands stemmed less from Christian perspectives on discrete and common subject material and more from the inclusion of content about the Christian story of God and his people. As pointed out earlier, Christian perspectives operated more at the level of the disciplines favoring a sociographic or cultural-geographic paradigm over an environmentalist and chorological one. At the level of individual topics, however, one, in general, looks in vain for Christian perspectives. Rather, one encounters materials related to Christianity not found in other sources. One common criticism found in reviews of general geography textbooks in Christian education journals is that materials related to church and Christian society are missing. Here is a rather typical example from a brief review of a geography textbook, *On a Journey through the Netherlands: Illustrated Geography Reader for the Public School*. After voicing criticism that this imaginary journey included visiting taverns and fairs, not appropriate settings for Christians, the reviewer (Wielemaker 1904, 176) wrote:

> The Christian school teacher also should not forget that books such as this need to be supplemented somewhat: the teacher also should take time to travel to Neerbosch with his students [Neerbosch was a

Christian orphanage settlement near Nijmegen], take the opportunity
to visit the museum of the Utrecht Missions Society in Utrecht, while
discussing the Veluwe [a unique physical region in the Netherlands]
take time to talk about the discovery of Hoenderloo by Dr. Heldring,
etc. etc. [*Heldring (1804-1876) was a Christian social activist who
founded many institutions of mercy including an agricultural colony
in Hoenderloo for neglected boys*].

When, as was often the case, Christian geography textbooks were not available,
not used, or unsuitable, the teacher, as described in this quotation, would be
expected to supplement the study materials with Christian content.

   After World War II, the perceived need and market for geography textbooks
in the Christian schools disappeared altogether; the focus of attention in
Christian education began to shift away from the subject material to the student,
pedagogy, and the educational community; moreover, secularization began to
make inroads, also in Christian education.[7] Recognizing that a geography
curriculum containing subjects especially for Christian schools was not likely to
be independently produced, a national Christian School Association in 1955
(Gereformeerd Schoolverband 1955) published a sixty-three page booklet in
which supplementary materials for geographic education in the Christian
elementary schools were outlined. Organized by continents, regions, and
countries and heavily slanted toward Dutch, European, and Reformed subjects,
this publication is a synopsis of religions throughout the world. Religions,
churches, and Christian missions are described for each region or country as well
as religious leaders and religiously significant places.

   Here is another example of augmenting the standard geography curriculum
with content of particular interest to Christian schools: the geographical
organization of religion in general, and of the Christian religion in particular. It
must be acknowledged, of course, and it is hardly surprising that this
supplementary content cannot be served up as a neutral inventory but comes
with an explicit Christian perspective: Non-Christian beliefs and practices are
filtered through and evaluated by a Christian worldview. In places, this booklet
also shows the dangers of misapplying a Christian perspective approach to
subject material: The perspective rules certain subjects as admissible and
significant, others as inadmissable or unimportant. For example, the entries on
the one page for Japan (Gereformeerd Schoolverband 1955, 28) comprise a
single paragraph on the principal religion, a blend of Shintoism and Buddhism,
a second paragraph on Christian missions, and the remaining half of the page is
devoted to Kagawa, the well-known Japanese Christian evangelist. Leaving
aside the not unimportant question of whether this treatment can in any way be
considered geographic, a more balanced application of a Christian perspective
would direct more energy to understanding Japanese religion from a Christian
perspective.

   Closer to home, the perspective in this publication becomes more sectarian.
The religious geography of the Netherlands is organized by provinces, with the
Catholic south virtually ignored and with the places of Protestant Christian

significance in the north thoroughly enumerated (Christian schools, hospitals for physical and mental illnesses, work colonies, institutions for alcoholics, difficult youth, the lonely, and so forth [Gereformeerd Schoolverband 1955, 44-55]). These faults aside, however, the purpose of this and other similar curriculum supplements is clear: to connect students to the Christian story and culture in each world region and country. A general principal for Christian geographic education is implicit in such content: When students in Christian schools study places around the world in their geography class, such learning must include the world of Christianity of these places. And in keeping with neo-Calvinism, the world of Christianity is not restricted to the church but embraces all of culture.

A Special Kind of Christian Content: The Geography of the Bible
The links in Christian schools between geography and the Bible were of two types: In the first, the Bible was a source of principles for the entire subject (previously discussed), and, in the second, geography served as supporting cast for the religion curriculum. In this second role, the geography of the Bible, of Palestine, and of the Near East in general takes on significance. The Bible as the record of the mighty acts of God and of human responses to His revelation was the foundation stone of the Christian schools, and knowledge of the Bible was an essential outcome for Christian education. The geographical setting of the Bible and of the lands of the Bible was regarded as an important assistant and enrichment for the subject of religion/Bible.

It is important to note that the geography of the Bible was generally taught as part of the religion curriculum and not as part of the geography program. In the world regional cycle of the geography curriculum there was more focused attention on Palestine than in public schools, but the geography of the Bible (lands) had its principal residence in the school's religion program. That made sense within the goal of imparting Bible knowledge. Geography's home, like other school subjects, however, was/is not in the religion program, underscoring the fact that the Christian character of geography is not to be narrowly identified with the geography of religion but is to be present throughout. Equally important, is that the field of geography of the Bible was and is not geography from the Bible or the Scriptures as textbook for geography. Even with the Bible laying the foundation for Christian education, Calvinist educators vigorously argued against a biblicism that looked on Scripture as the source of specialized scholarly knowledge; instead, the whole of God's creation was text for geography. Again, the geography of the Bible could not take sail without the pilots of a neo-Calvinist or some other perspective.

Van Deursen (1927-28, 406), enthusiast and specialist in the geography of the Bible and of Palestine, summarized these issues beginning with an unreferenced quotation from George Adam Smith, the well-known author of *The Historical Geography of the Holy Land* about the purpose of the geography of the Bible:

> . . .[I]t is the love for the Holy Bible, the longing to more closely know the land which is the stage of sacred history. And that is why

we teach the geography of the Bible. Of course not in the meaning of pietism, which would happily see all instruction wrapped up in religious education and therefore would most like to see geography taught from the Bible. Nor as Melanchton, who only found geography necessary, in order that we may know where God revealed Himself and in what places He showed His works. Such a restriction would be against the meaning of Scriptures, where we listen to the voices of the angels: *the whole earth* is full of His glory. (Isaiah 6:3) But even if we acknowledge the glory of God, which reveals itself in all places of His dominion—nevertheless the territory of Special Revelation is the land which because of our religious consciousness has the deepest interest.[8]

The geography of the Bible was one important layer of Christian content for geography in Christian schools even though it was largely taught in religion classes by nongeographers. Indeed, because, as we have observed, preparing more general geography materials from a Christian perspective was very daunting, without much precedent, and not seen as essential by many; if one wanted to be busy as a Christian educator-scholar in geography, then the geography of religion and of the Bible and the Holy Land was a self-evident choice.

The life and work of Arie Van Deursen, lifelong Christian high school geography teacher and scholar, is an example of such a professional orientation in geography. As we have already observed, Van Deursen made many significant contributions to Christian geographic education and scholarship, including reviewing countless geographical publications in Christian education journals and magazines, publishing several influential Christian perspective pieces for geographic education, and coauthoring a five-volume series of geography textbooks for Christian high schools. Yet, he poured the lion's share of his prodigious scholarly energies into the geography of the Bible and of Palestine and was best known and respected for that work. This overriding scholarly interest produced the largest number of book-length accounts, including overviews of biblical archaeology, atlases for the school subject of biblical history, several geographies of the land of the Bible, a set of large classroom illustrations about the land of Palestine used in Bible courses, and several Bible reference books (Mulder 1988; Smilde 1963-64).[9] Other European geographers might consider the Greco-Roman world the spiritual homeland of the West; Van Deursen looked to Palestine, the land of the Bible. Can a spiritual heir forget his homeland?

Conclusions

The development of geographic education in Christian schools in the Netherlands was the largest, most organized, and most significant historical episode to culturally construct geography by the contours of a Christian worldview. Not typically confined to the academy and scholarship, it reached into society at large. One way or another, this construction of geography affected the learning experience of millions of students and touched the professional lives of tens of thousands of teachers. It was authored by and comfortably fit into the cultural milieu of Dutch Reformed culture.

Christian geography educators spearheaded the movement into a sociographic paradigm bringing their own distinctive rationale to the successful effort to replace the place name/relief model of geographic education in the Netherlands. This paradigm provided room for the kind of content that Christian education considered strategic. While other communities of geography educators and scholars moved into a similar direction for their own reasons, sociography constructed by Christian geographers continued to show some distinctions, notably in the content areas of religion and culture. Sociography cannot be doubted that the widespread adoption of this paradigm represented true progress when compared with earlier geographic education. At the same time, it must be emphasized that Christian geographers bent the sociographic approach to fit their own interests. It did not originate in the Christian academic and educational community as a product of pioneering systematic work and collective Christian reflection but, rather, came fully assembled from elsewhere. As a secular school, sociography breathed a spirit of descriptive empiricism and cultural relativism. Its deep structure contained no Christian categories.

Geography as sociography would today be seen as a social-studies curriculum. The challenge for such a school subject, then and now, is how to integrate very diverse materials into the curriculum and into the classroom. While Christian geographers pointed to religion, world and life view, and way of life as integrating themes, in practice, much of geography, structured by a sociographic paradigm, remained an omnium-gatherum with religion in its own compartment; such integration is indeed very complex and daunting. Nevertheless, well-integrated social-studies units with a Christian perspective are attainable. Beginning in the 1970s, the Institute for Christian Studies in Toronto, a Reformed academic institution, established the Joy in Learning Curriculum Development and Training Center. It prepared, published, and distributed a number of integrated social-studies curricula (De Graaff, et al. 1980; 1981). However, if education in the Netherlands had stuck with more mainstream geographic paradigms, such as human landscape or society-environment, there would have been opportunity to construct a distinctive Christian approach to geography with more scope and focus and greater ease than such integrated social studies.

The general characteristics of a neo-Calvinist worldview contain more possibilities for constructing a Christian geography than the scriptural principles delineated by Van Deursen and others. The general features, themselves tethered to God's special and general revelation, have served as a basis for the development of a Christian orientation, of theorizing, and of systematization in other fields, for example, philosophy and aesthetics (Zuidervaart and Luttikhuizen 1995; Griffioen and Balk 1995). Moreover, they have served as keys to discover, interrogate, and evaluate the histories, worldviews, and paradigms of academic disciplines. When, as they have, Christian educators sense and document a spirit of materialism, socialism, and intellectualism in the geographic work they were reviewing, then a neo-Calvinist worldview is at work. The scriptural principles, however, appear somewhat out of place within neo-Calvinism, bringing the biblical text too close to specialized knowledge; additionally, their relevance for the praxis of geography remains obscure.

Many neo-Calvinist geography educators in the Netherlands acknowledged that by its very nature a Christian worldview should differentiate the treatment and interpretation of geographic subject material across the board; in practice, however, they mainly selected and added geographic stock related to the Christian story such as missions, church, institutions of mercy, and the geography of the Bible and of Palestine; these were usually missing from geography textbooks. The larger conviction remained unfulfilled because of a lack of institutional and individual leadership in the development of an alternate paradigm. If the kinds and level of resources, theoretic insights, strong commitments, keen intellects, momentum, and following that were part of the creation and institutionalization of the sociography paradigm would have been present at the Free University during the first half of this century, a critical mass for the development of a neo-Calvinist geography paradigm would have been present. It was not, and Christian geography could not grow up.

Notes

[1]This chapter is part of larger research program related to Christian geographic education and scholarship in the Netherlands in the twentieth century and is sponsored by the Calvin Center for Christian Scholarship. I am grateful to the Calvin Center for providing financial support for two extended research trips to the Netherlands during the summers of 1994 and 1995. These provided opportunity to collect, photocopy, and read many of the primary sources that are analyzed in this paper. I am also grateful to the philosophy department at the Free University of Amsterdam, and to its chairman, Sander Griffioen, for providing facilities as well as warm personal and academic hospitality.

[2]This quotation is attributed by Romkes to Steinmetz but is unreferenced.

[3]I am indebted to G. A. Hoekveld for this observation.

[4]Consult the special history of the Free University written to mark the occasion of the centennial of its founding (Van Os and Wieringa 1980).

[5]Although part of the larger research project on geography in Christian schools in the Netherlands, a more complete analysis of these textbooks is beyond the scope of this chapter. It will be important to compare similar types of textbook materials published for public, Christian, and Roman Catholic schools.

[6] VerMooten and Sleumer's (1929-30) views of geography textbooks for Christian schools and Christian geography are set out in a series of articles in the journal *Christelijk Middelbaar Onderwijs* [*Christian Secondary Education*]. These brief articles are a set of serial responses to their critics following their initial article about the Van Deursen et al., five-volume high school geography textbook series, *De Wereld en Die daarin Wonen* [*The World and Those Who Live in It*]. These need to be read in conjunction with the intervening articles of their critics.

[7] In this regard, it is instructive to consult the geography section of a handbook for Christian education published in the early 1950s. (Van Hulst, et al. 1952, 168-206).

[8]This entire two-part article is a useful overview of the geography of the Bible in neo-Calvinist education. In distinction from environmentalist explanations/reductions of Scripture, the Bible remains the authoritative source, and geography provides the physical and cultural milieu of the accounts, thereby enriching them.

[9] Two brief biographical sketches of Van Deursen have appeared, one by H. Smilde (1964-65) shortly after his death in 1963, and the other, by H. Mulder (1988).

Chapter 8

Perspectives, Worldviews, Structures

Sander Griffioen

Perspectives

AS ANNOUNCED IN the Introduction we will now take a broader look at the worldview theme. This means that our investigation does not start from within the domain of the special disciplines in order to look for broader and deeper implications *of* what scholars actually do. Nor does it, strictly speaking, follow the other approach outlined in the Introduction: It does not start from a Christian understanding of our world in order to trace implications *for* geography and other disciplines (although tacitly such an understanding is certainly present). Rather, the investigation's first interest is to assess the opportunities that the present climate of thought offers to the pursuit of Christian scholarship. There is a real need for such an evaluation. In the previous chapters, much has been said about the collapse of the positivist paradigm and the remarkable tolerance for diversity that has come in its wake. On the whole, the tone about this development has been positive, albeit mixed with concern about attendant relativist tendencies. Yet, thus far, an overall assessment is missing. Briefly, the question is: What does the abandonment of the once dominant paradigm mean if related to worldviews? How open is the postpositivist climate of thought to acknowledging the role of worldviews?

Worldviews belong to the general class of perspectives. These may be no more than ways of perceiving reality. A worldview implies a way of seeing but also includes other characteristics. The notion of worldview adds (at least) two features to the general semantics of the word *perspective*. First, it not only represents ways of looking but also implies certain beliefs about the world. There is more at stake between worldviews than just perceiving one and the same thing from different standpoints: Rather, it is a matter of different beliefs about the thing itself. Second, these beliefs are likely to be in conflict. Different views on one and the same thing do not need to be conflicting: They may well complement one another; however, once we allow for different beliefs about the same thing, then the nature of the thing itself is at stake. It is then no longer certain that the views complement each other. Rather, given the comprehensive nature of worldviews, it is safe to consider conflict among beliefs as inevitable.

There are two pitfalls to avoid here. On the one hand, there is a tendency to merely take worldviews as ways of seeing the world. On the other hand, conflicting beliefs nowadays are often interpreted as incommensurable (i.e., as having no common measure), and the resulting conflicts are regarded as unresolvable, blind conflicts without winners or losers. A characteristic of the present climate of thought, whether we call it postmodern or not, is to do both. Worldviews are often reduced to mere perspectives. At the same time, when beliefs are taken into account, they tend to be conceived as incommensurable. Both attitudes are present in the current debates about multiculturalism. We encounter a fascination with cultural perpectives next to a fatalistic interpretation of conflicts between cultural groups.

I do not want to belittle the openess toward perpectival differences as such. It is good to be aware of the fact that different people view the (their) world differently. Scientists see reality different from let us say, cab drivers, and so on. One of the exciting things today is a heightened awarenes of the role that cultural factors play in perception. By way of example: Arabian people perceive water differently from the Chinese for whom control of the rivers has been a matter of life or death (so much so that they use the same character for to govern as for to control water [*zhi*]), and differently again from the Dutch for whom, as Simon Schama reminds us in his book, *The Embarrassment of Riches* (1987), on the one hand the sea posed a perennial threat to their habitat behind the dikes, while on the other hand it represented a source of wealth with all the possibilities of overseas commerce.

There is much openess, if not fascination, with respect to cultural differences of this kind. However, there is more bite to worldviews than to (mere) perspectives. As will be explained below, the metaphor of mapping aptly characterizes the nature of beliefs about the world. Worldviews (pretend to) map the basic features of a given domain and thus show what roads to take or not to take. This metaphor helps us to understand the nature of the truth claims involved: A worldview pretends to offer a reliable map. Conflicts then are not to be understood irrationalistically as collisions between incommensurable convictions but as rival claims with respect to identifiable issues.

Pushing the map metaphor further, a major difference between two kinds of worldviews comes to the fore. On the one hand, we find those that largely remain implicit to the practice of the sciences, such as tacit assumptions about the nature of the universe, and so forth. Worldviews of this kind are traditionally designated by the German word *Weltbild* (world picture). On the other hand, we find those having a normative, often programmatic function, the so-called *Weltanschauung*. Although a *Weltbild* may be very unemphatic, such as when we speak of the world pictures of the authors of the Bible, yet, on the whole, my position is that worldviews of this kind do fulfill an orientational role that may not be innocent. For example, there is abundant evidence that the so-called scientific worldview has played a major role in the secularization of the West: Through the application of scientific results, the implict *Weltbild* has found new roots in everyday life at the expense of traditional religions.

These two kinds of worldviews may also be interconnected. There are many cases in which a *Weltbild* received a normative role and thus started to function as a fullfledged *Weltanschauung*. This is what has often happened with the so-called scientific worldviews. Illustrations include the mechanistic worldview of early modern times, the positivism about which I spoke earlier, as well as present-day cosmologies such as Prigogine's order out of chaos and Stephen Hawking's history of time. Down-to-earth scientists often react to such speculations by deciding to stay away from worldviews altogether. However, the thrust of this book is different. As the Introduction has suggested, the challenge is to formulate a normative worldview that is able to do justice to, and disclose the inherent potentialities of, the special disciplines.

After the introduction of the *Weltbild-Weltanschauung* distinction, a further differentiation will be made between worldviews of social movements and those geared to scholarship. Then Habermas's contention that the social sciences increasingly abstain from global interpretations of nature and history will be discussed. This will lead to a comparison between Weber and Kuyper, followed by a discussion of Wolterstorff's critique of religious totalism. My answer to the problem raised by Wolterstorff follows the direction hinted at in the Introduction: to combine the normative with the implicit, the structures *for* and the structures *of*.

Openings

I will start by taking religion as a touchstone to assess changes in the climate of thought. The scientific mindset has long been dominated by an evolutionary positivism. Basically it was a matter of either-or: Either one followed religion, or one followed science. The former was viewed as a kind of proto–science, bound to disappear once genuine science entered the scene. Marx's view of technology offers a graphic illustration: Technology stands for real mastery over nature; religion stands for illusionary mastery along with magic and mythology. The advent of the first, ends the reign of the latter.[1] In intellectual circles, however, it was positivism rather than Marxism that represented this kind of thinking. David Livingstone's chapter gives some telling illustrations of the positivist mindset in geography. The great ideal was a unified science: a body of theories built on a rock bottom of objective knowledge. Worldviews (along with other subjective factors) were, at best, given a secondary role as a projection or extrapolation of scientific data, constituting the so-called scientific worldview I alluded to earlier. The situation has now changed drastically. The ideal of objective, unified knowledge has lost its vigor. It is not uncommon nowadays to hear scholars of different stripes and colors publicly declare that their own work has been guided all along by broader visions of what the world is like and, even, by what it is that makes human life meaningful. Although such confessions do not automatically imply a rehabilitation of religion in scholarship, the easy dismissal of religion as pertaining to a prescientific era now sounds like an echo from a past era. Only fervent atheists still view science and technology as substitutes for religious convictions, but they no longer set the tone. Put positively, there is a growing willingness to take worldview issues seriously, even when these are approached from an explicitly Christian standpoint.

A few years ago a Dutch physicist, Arie Van den Beukel, published a book in which he strongly defended a Christian understanding of reality over against several varieties of the scientific worldview. To his own surprise, this publication has gone through reprint after reprint. On various occasions Van den Beukel has admitted that he himself had been struck by the attitude of his colleages at the Technological University of Delft: Whereas he had anticipated that the book might damage his reputation as a serious scientist, in fact, he received encouragement to go on with this work.

Also noticeable is a growing acknowledgment of pluralism as a normal state. It is received wisdom nowadays that any subject allows for a plurality of perspectives. If it is true that worldview approaches are presently on the ascendency, then, significantly, only in plural forms. Today there is not any standard worldview. This is almost common sense. But why associate such plurality with openness? It is because the acknowledgment of plurality liberates us from the obsession that consensus must be reached at any cost. It should be borne in mind that the ideal of unified science was built on precisely this assumption: Science fosters consensus. Religion divides, science unites. Where did its power to overcome division originate? The answer, of course, was that objective knowledge is indubitable knowledge, freed from volatile opinions and divisive creeds. We should never forget that only on these premises could science have become a trump in the hands of the Enlightenment thinkers. Only if this connection is grasped does it become possible to fathom how decisive the acknowledgment of pluralism is. It is the very plurality of perpectives that contradicts the deepest aspirations of the Enlightenment.

Closures

Let us now briefly consider the closures stemming from the present climate of thought. Postmodern thought resembles pagan gods that take with one hand what they have given with the other. It liberates scholarship from the obsession with certainty, but it fails to show a way leading beyond present pluralities and thus leads into a new form of captivity. Its merit is to have defended a narrative mode of reasoning and by doing so it opened up the theoretical discourse to the quest for meaning. Yet, simultaneously, postmodern thought opened the gates wide to a relativism with respect to any truth claims made on behalf of such narratives. Significantly enough, incredulity toward metanarratives (Lyotard's formula) became the standard definition of postmodernity. This incredulity expresses just as much a refusal as an inability to believe. Postmodern thinkers refuse to place any hope on the grand syntheses of the past, the grand projects about which David Ley speaks in his chapter. They know only too well what the costs of the pursuit of unity have been: the marginalization or abandonment of all elements that did not fit into the picture, from human emotions, the body, and women, to the poor at the periphery of the world system. All this is well known. What remains mostly hidden to the public eye is that this incredulity also issues from a deeply agnostic attitude toward truth claims of the metanarratives. Characteristically, the metaphor of the nomad has recently gained currency as a key to understanding the postmodern ethos. It is the image of nomads trekking from tradition to tradition, without committing themselves to any story in particular. In this way, traditions become options instead of being lived; the narratives become mere stories, without claims to truth and allegiance reaching beyond the subjective.

Metanarrative is not in all respects an apt phrase. It is a portmanteau term covering different elements that a more careful analysis would want to distinguish. A given culture may be analyzed in terms of a shared history, customs, institutional patterns, values, and worldviews, but last but not least, it is a matter of beliefs. A narrative approach has the potential of including all these factors. To that extent, it drives home the point that perspectives do not stand on their own but are always part of a greater whole. If isolated, perspectives become mere schemes adopted or abandoned at one's pleasure. What is missing here is the relationship of such perspectives to specific beliefs about the world. The word belief should be taken in an emphatic sense. To adopt a worldview implies an assent, an amen so to say, an affirmation, be it tacit or open, that this indeed is what the world is like, that this is how different domains of life cohere. I submit that this fiduciary aspect belongs to worldview itself, instead of to something that only pertains to certain ways in which worldviews are used. When the inner assent is missing, one ends up with a manifold of fungible perspectives. What then emerges is a situation that can be characterized both as relativism and perspectivalism. Relativism because of the fungible character of the perspectives that leave a person without an intrinsic reason to prefer the one over perspectivalism because the individual perspective turns into a windowless monad.[2]

Different Maps

To see sharply what is involved in the beliefs about the world to which I have referred several times, we do well to return to the definition of a worldview given in the Introduction: "Their worldview is their picture of the way things in sheer actuality are, their concept of nature, of self, of society. It contains their most comprehensive ideas of order." (Geertz 1973, 127) Note the reference to order at the end of the quotation. It is on this basis that Geertz distinguishes worldview from ethos and values: The former relates to order and structure instead of expressing just a tone, a quality, or something similar (129). Whereas the elements just mentioned are evaluative rather than cognitive, a worldview expresses basic beliefs about the order of reality. The obvious metaphor to render the essential quality of Geertz's definition is that of a map. Worldviews map out domains of reality. The fact that they do so in a global, rather than in a detailed way, does not turn them into mere perspectives. Any worldview will pretend to offer a reliable map of the world. In an increasingly complex world, the first need of people is to have a global map that shows general features that are needed for orientation.

Obviously these maps are not all the same. Some offer clues about the riddles of the universe: how heaven and earth came about; what the place is of the human person in the cosmos; the struggle between good and evil, life and death. Others concentrate on the nearby human life world, the universe next door (Sire 1988), rather than on a cosmos far away. The reason for the shift is, I believe, the rising complexities of human society, in which, on the one hand, the forces of modernity all but wiped out the traditional boundaries while, on the other hand, they stimulated a process of differentiation of societal spheres and thus created many new boundaries. More than before, clues were needed about how such societal spheres cohere, how to live out of basic convictions in different areas, how to maintain a consistent lifestyle, and what norms there are to live by.

For our purpose, it is important to distinguish two types of maps. On the one hand, there are worldviews normally unreflectively held, uncovered, excavated by anthropologists and other scientists. The quoted definition of Clifford Geertz refers to this type of map. Worldviews of this kind are passed on from generation to generation by way of nurture, ceremony, and storytelling rather than formulated and discussed as such. Hence, maps of this kind do not possess the cognitive clarity of typical exemplars of the second species that I want to distinguish, viz explicit and more or less worked out worldviews.[3] It is only here, in relation to the second type, that it makes sense to speak of convictions as well as of people's self-consciously taking a particular stance. In some languages (German and Dutch, for instance) the distinction between these types corresponds to two distinct concepts. The typical German word for the first type is *Weltbild* (world picture), while the second is rendered by *Weltanschauung* (worldview). In practice, the distinctions are not always clear-cut. Some thinkers are obstinate enough to prefer the one where we would prefer the other; a case in point is Heidegger, who, in his famous essay on the era of the worldview, prefers *Weltbild*, while in fact *Weltanschauung* more aptly characterizes the modern era.[4] Of course, there are also authors who do not use either one, preferring synonyms instead. A case in point is Voegelin's distinction between noetic and nonnoetic self-interpretations (*noetic* derives from the Greek *nous*, i.e., mind). The latter formulations fit well with the *Weltbild* concept as delineated above. Voegelin wants us to think of symbols through which a society expresses its experience of order. Those symbols may be mythical, as in traditional societies, or revelatory, as was the case with Israel, or ideological, as in modern times. All that these symbols have in common is that they are not self-transparent: "There are no societies whose constitutive self-interpretation is noetic." (Voegelin 1990, 144) Conversely, noetic interpretations (read: *Weltanschauungen*) do imply a level of articulateness only found with visionary individuals and the communities inspired by them.

The distinction between these two types of maps becomes even more interesting if the differences with respect to cultural pluralism are taken into account. As a rule, worldviews in the sense of *Weltbild* present themselves in a diachronical order: We speak of the typical world picture of Medieval people's making place in due time for that of the Renaissance. Although these pictures are at odds with one another, conflicts become palpable only at times of transition and boundary clashes between civilizations. An illustration of this kind is the collision between Eastern and Western concepts of space when modern physics was introduced into the Oriental world. Needham, in his influential series on *Science and Civilization in China*, has shown that the Chinese scholars, for all their subtle qualitative distinctions, had no equivalent for the Newtonian concept of homogeneous space.[5] Interestingly, as a consequence of Japanese modernization in the 1870s and 1880s, a similar clash of perceptions and representations of space and borders developed between China and Japan (Howland 1996). With regard to *Weltanschauung*, conflicts come to the fore more readily. Not only within one culture or epoch do we always find a plurality of worldviews, but also, because they are explicit, the conflicts easily come into the open. Wilhelm Dilthey (1833-1911), one of the first to raise the concept of worldview to philosophical dignity, poignantly formulated what the introduction of worldviews into the public discourse meant: a *Kampf der Weltanschauungen*, an open battle at the core of modern culture.

It goes without saying that it makes quite a difference whether the central theme of this volume, *Geography and Worldview*, is taken in the one sense or the other. The focus may shift accordingly from broadly shared assumptions about the cosmos, nature, the human world, and natural theology to the coexistence of rival schools of geographical thought. Especially the previous three chapters (Aay, Hoekveld, Hoekveld-Meijer) underscore the necessity for geographers to reflect on the distinction between these two types of worldviews.

Worldviews in Theory and Practice

Yet another distinction within the worldview idea is called for: those with the social and political arena as home base and others primarily at home in the world of scholarship. This distinction does not run parallel to that of *Weltbild-Weltanschauung*. While sociopolitical worldviews characteristically are of the *Weltanschauung* variety, one meets both types in the world of scholarship. As we have already seen, worldviews function both implicitly and normatively within the academy.

The most formative influence on *Weltanschauung* have been the great social movements. Within the whole gamut from Communism to conservativism, worldviews were nourished not just for propaganda purposes but also because of a genuine need for orientation. Dilthey's motto about the struggle of the worldviews refers first of all to exemplars of this type, i.e., the idealist outlook of socialism-Marxist dialectical materialism, the naturalism of the social-Darwinians, the mechanistic outlook of laissez faire liberalism. However, Dilthey also actively participated in the debates about theory that flared up in his days. In Kant's era, the word *science* was still univocal enough to be used without further qualification. A century later, this was no longer the case. A need for regulative and integrative ideas and visions had arisen as a consequence of an ongoing differentiation of the sciences. The presence of the social sciences—humanities—next to the natural sciences—mathematics and biology—created problems for the demarcation of fields of study as well for the scientific method. That these problems are still with us today need not worry us: Fundamental issues cannot be decided by scientific arguments and therefore remain untouched by scientific progress. Dilthey was one of the first to acknowledge the lasting significance of worldviews in this respect.

It is unfortunate that so often the two varieties here distinguished are lumped together. Lyotard (1984) did so in his influential book, *The Postmodern Condition*, and although, strictly speaking, his arguments pertain to sociopolitical worldviews only, it has become common to assume that metanarratives are on their way out. Also, if we accept Lyotard's point (which to some extent I do[6]), it still remains uncertain whether the same would hold for worldviews operative within scholarship. This is not to deny interconnections between the worlds of science and of social movements. Take the modern idea of progress: Undoubtedly, faith in scientific progress was greatly stimulated by the emancipatory elan of new social groupings, bourgeoisie and working class alike. Yet, on closer scrutiny one sees that the idea of social progress only offered a global orientation to the scientific enterprise. In fact, it allowed for quite different (and conflicting) elaborations, ranging from Darwinian naturalism to Kantian moralism. Neither socialism nor liberalism (or conservatism for that matter) have been able to offer distinct regulative ideas for the social sciences.

But, in fact, do the sciences readily show the presence of full-fledged worldviews? In the daily practice of scholars, would one not find congeries of assumptions, hypotheses, schemes, and ideas borrowed from others and put together for limited purposes rather than grand stories and views of the world? Is metanarrative at all an adequate designation for an average theory? No one less than Habermas (1982, 584), not known for postmodern sympathies, has stated that sciences should more and more abstain from global interpretations of nature and history. Would geography be an exception?

To answer these objections, I begin by addressing a misunderstanding. The geography and worldview theme does not mean that complete worldviews are patently present in any context, just waiting there for us to be analyzed. The worldviews dealt with in this volume as a rule do not surface but only become visible through our labor of extending and extrapolating the arguments offered; this requires imaginative interpretation. Put differently, analysis of worldviews is never just a matter of observation, but always (also) of imaginative interpretation; what is called for is a diagnosis reaching beyond the immediately given.

Notwithstanding, we should not forget that grand theories are still around. Their popularity may even be on the rise if Quentin Skinner (1985) is right. Moreover, since he published his book *The Return of Grand Theory*, other versions have begun to attract attention. A case in point is the work of William McNeill, 1996 recipient of the prestigious *Praemium Erasmianum*. His most recent work, *Keeping Together in Time* (1995), is organized around a relatively simple idea: Dance and drill as analogous means to promote social cohesion. But it soon becomes apparent that the organizing theme is embedded in a grand theory covering the whole of human evolution, starting with dancing primates and ending with the need for dance (and drill?) in the multicultural society of the 1990s. Thus, the book offers a nice example of a simple theme filled to the brim with worldviewish implications.

Truth Claims

The previous example serves to illustrate another point. Worldviews do not just represent empirical tendencies, but instead, are normatively laden: They harbor claims. McNeill's worldview pivots around the role of social cohesion in evolution. Should the pivot be qualified as a hypothesis or as a basic conviction? I think it is both. To start with, it is clear that a strong conviction is at work. It seems fair to say that the assumed relationship between dance and drill was not chosen as an arbitrary starting point, as just one hypothesis among other equally plausible hypotheses. Apparently, the author went on his way trustingly, confident that the approach taken in his previous works would yield good fruits once more. It is this trust that I described earlier as a fiduciary aspect (distinguishing worldviews from mere perspectives). Nevertheless, all the characteristics mentioned thus far do not turn a worldview into a confession of faith. In this case, when the project started, the author, of course, did not know for sure that the dance and drill assumption would indeed work as an organizing idea for the historic materials at hand. Put differently, it is always possible that a worldview fails to illuminate and organize a given subject matter. Worldviews arc trees that must be judged by their fruits. If they fail as a regulative and integrative idea, they no longer deserve to be taken seriously. Here then, finally, I have identified the spot where Lyotard's incredulity verdict applies. Illustrations of failing metanarratives are not hard to find. How could it be that neo-Marxism has left so few traces after having triumphantly entered the universities (and geography) less than three decades ago? Should one not conclude that it has failed to live up to its claim to radically transform the scientific enterprise? And what happened to the grand schemes of so many other trends that were once fashionable?

If it is true that worldviews come with claims, then it is also true that they are competing among themselves about very real things. This point has not always been clear. Often the clash of worldviews was associated with the idea of an ongoing, never-ending battle between incommensurable perspectives. Max Weber (1864-1920) is responsible for this unfortunate suggestion. He presented the struggle between the worldviews as a spiritual battle, the outcome of which could only be decided by fate. Now that Christianity was losing its hold, he thought that the eternal struggle between the old pagan gods was due to resume with the new gods, the ideals of the great social and political movements:

> Many old gods ascend from their graves; they are disenchanted
> and hence take the forms of impersonal forces. They strive to gain
> power over our lives, and again they resume their eternal struggle
> with one another (Weber 1994, 17).

There was a side to Weber that welcomed this battle, yet he also defended the neutrality of science. He saw political life as a battle of spirits; here he left no place for neutrality. Yet, as a scholar he wanted very much to keep the worldviews outside the gates of the university for fear that, otherwise, science would lose its authority. Thus, with respect to the social sciences, he kept to an objectivity of the strictest sort. A national figure at the end of the First World War, someone to whom students looked amidst the breakdown of traditional values, he banned all worldviewish subjects from his lectures, claiming that scientific objectivity demands that professors abstain from offering recipes for the good life. The social sciences, he stipulated, must ascetically abstain from value judgments. Only then would they be able to play a role as impartial arbiters.

Weber's dual attitude toward pluralism represents a kind of schizophrenia that is all too common in the universities. To the many Weberians still around, the theme of this book, geography and worldview, can only signal a fateful desire to surrender the discipline as a free academic pursuit to the powers that be. But, we must ask, is not the problem rather that Weber and his followers are unable to conceive of worldviews as hypotheses claiming truth? In Weber's view, the only claims in this respect are claims to power. Granted, the power aspect of worldviews cannot be ignored: Weber has clearly seen that it is the nature of worldviews to strive to gain power over our lives. But could they yield power without the assent of humans? And how would they gain consent without using arguments that may or may not convince?

At stake in this discussion is whether worldviews are, at depth, blind powers falling under the same rubric as myths and ideologies, or, to the contrary, powers that are only able to enslave by appealing to arguments for certain states of affairs. The second position is Kuyper's and his followers. In the Introduction, Abraham Kuyper (1837-1920) was introduced as a postmodern thinker ahead of his time. In a certain sense Max Weber would be a better candidate for this title. Interestingly enough, when Kuyper had become prime minister, Weber followed news about his university reforms with considerable interest (Griffioen 1994, 25). Yet, Weber's own idea of an inconclusive battle comes closer to postmodern perspectivalism than to anything Kuyper stood for. The latter understood the worldview battle as a contest with winners and losers. He placed high hopes on the power of the Christian worldview to win the race. What made him so confident? He was convinced that only a Christian worldview opens a meaningful perspective on the whole of reality. The competition then is between totality views, the great - isms: Christian worldviews (Calvinism and Romanism), modernism, and older and newer forms of paganism. Kuyper's is a truly grand vision encompassing not just theology and philosophy but also the sciences and even culture at large. At the center stands his confession that not "one square inch" of reality is exempted from the struggle between the kingdom of darkness and the kingdom of Christ. The influence of this vision has reached far beyond the milieu of (neo-)Calvinism. George Marsden may not have exaggerated when he spoke of "the triumph—or nearly so—of what may be loosely called Kuyperian presuppositionalism in the evangelical community."[7]

Religious Totalism

It must be admitted, though, that Kuyper's emphasis on totality views was not without risks. In a certain sense, the tradition in which both Calvin College and the Free University stand produced its own kind of perspectivalism. It is important to pay attention to Wolterstorff's critique of what he calls religious totalism. Nicholas Wolterstorff belongs to the generation of students brought up at Calvin College with Kuyper's *Lectures on Calvinism*, the Stone Lectures, delivered at Princeton in 1898. He describes Kuyper's thesis that scholarship is not religiously neutral as "one of the most insistent and provocative claims of the neo-Calvinist movement" (Wolterstorff 1989, 56). Although positive about Kuyper's repudiation of the consensus ideal of the unity of science, he disagrees strongly with the way in which the relation between faith and learning is conceived. Kuyper held that the great divide between faith and unbelief must lead to a similar bifurcation in science. Put succinctly, in Kuyper's words: "two kinds of people, hence two kinds of science" (66). Wolterstorff rejects this thesis as a form of religious totalism (63-66). He does not want to deny the possibility of deep contrasts between beliefs, but only in the sense of aposterori disagreements, whereas Kuyper grants differences an apriori character (66-67). From his essay, it becomes clear that Wolterstorff prefers the generic notion of beliefs to worldviews, apparently because he holds that totalist seduction inheres in the very concept of worldview itself.

Wolterstorff must be credited with having identified a weak spot. If considered in isolation, the notion of worldview seems to imply that principles (and similar apriori elements) always strive for complete self-realization and thus for the fullest possible expression of their mutual differences. Obviously, once divergence receives so much emphasis, it is only a small step to a perspectivalism that, as mentioned before, makes perspectives become total and hence incommensurable in relation to one another. On comparing Weber and Kuyper, it turned out that the latter did not yield to the irrationalist notion of a blind struggle between worldviews but, instead, held on to an open contest in a public arena. Kuyper, however, was less successful in philosophically clarifying the stakes of such a contest. As indicated, God's creation is the horizon against which worldviews are placed. What is lacking is a clear view of the whole fabric of creational structures and norms, in which respects, perspectives agree and disagree, therefore, remains in the dark. Theologically, Kuyper sought to balance antithesis between belief and unbelief and communality between believers and unbelievers by grounding the latter in common grace. Philosophically, however, his way to account for the possibility of communication with nonbelieving scholars was by allowing for exceptions to the two-humanities-two- sciences rule. He held that "the workings of our senses" as well as "the formal process of thought" remain untouched by unbelief or faith. Thus, he exempted large parts of the empirical and formal sciences from the two-humanities rule (Wolterstorff 1989, 60). In this way, ironically, a kind of religious neutrality was allowed to slip in through the back door.

It is a lesson for us that in recent decennia new arguments against neutrality have come precisely from domains exempted by Kuyper. Thomas Kuhn's paradigm theory pivots around the thesis that perception is theory laden: He has clearly shown that sensory perception depends to some degree on orientations that people follow. In the arena of formal thought processes, it may suffice to refer to Alvin Plantinga's work on modal logic, which shows clearly that precise logical reasoning is based on assumptions.

Directions and Contexts

In the meantime, it has become clear that it is one thing to detect worldviews that are operative within the sciences but quite another to properly assess such findings. How are worldview clashes to be interpreted? How are conflict and communality between interpretations distributed? At this juncture, it helps to distinguish among *directions*, *contexts* and *structures* (see Mouw and Griffioen, 1993). Let us start with directions. Worldview is a sensitive instrument to register directional conflicts. Yet, as will become apparent, it is of less help with respect to the other dimensions.

Max Weber was keenly aware of the relationship between worldview and direction. His study of the social ethics of the world religions offers a graphic formulation of the function of a worldview. After defending the notion that human action is determined by interests rather than by mere ideas, he goes on to qualify his position as follows:

> However, the worldviews, which have been created through ideas, have often determined—as do rail-attendants—the tracks on which the dynamics of interests move actions forward. Indeed, the

worldview determined from what and for which one wanted—and could be—redeemed (Weber 1922, 252).

Although both material and ideal interests work as propelling forces, the tracks followed are determined by something else: Weber's "from what" (*wovon*) and "for which" (*wozu*) are worldview matters.

Weber's formulations catch directionality quite well. He has understood that visions on reality always answer questions as to the "from what" and "for which," either implicitly or explicitly. His point is that conflicts between social ethics should not be seen primarily as clashes of interests, as Marxists do, but as diverging views as to origins and destinations. I think the same holds for the relation between theories. An illustration with ample relevance for geography would be the divide between positivism on the one hand and hermeneutical theory on the other. Whereas positivism (as well as much of analytical philosophy) betrays a nostalgia for a pristine reality untouched by human interpretation, hermeneutical theory (along with interpretative sociology) attribute little less than world-formative powers to interpretation: Here the mistrust of interpretation (subjective, prejudiced) has made place for a celebration of the disclosing powers of human understanding.[8] A landmark in this respect is Gadamer's rehabilitation of prejudice as an indispensable beginning for the process of knowledge (Shin 1994, 74-5).

Since Weber, awareness of directional divergence has had an uneven development. The renaissance of Marxism in the 1960s brought a heightened awareness of conflicts between visions and thus theories. Since then concerns have shifted toward issues of contextuality. At the same time, complementary relations between theories have drawn more interest than root conflicts. Context in this connection refers to the great variety of circumstances in which people live. As an old sage said of human beings: "Their natures are much the same, their habits become widely different" (Wang Yin Ling, 1935).[9] If in scholarly work it is agreed that their range of application theories ought to be as context specific as possible, then it is also natural to assume that theories do not contradict if properly contextualized. Thus, the relation between theories is viewed as primarily complementary: A theory having been confirmed within a certain range of situations will need to be complemented by other approaches outside its own domain. Conflicts, if acknowledged at all, are likely to be attribted to a lack of specificity rather than to directional diversity.

The concern with contextuality is certainly valid. Undeniably, much harm has been done by a quasi-universal application of theories. This is true, for instance, of modernization theories: Based on Western experience, these have often been applied blindly to non-Western societies. Thus, it was commonly overlooked that modern secularization is a typical Western phenomenon. Consequently, with few exceptions, social scientists working with a modernization concept have failed to predict new religious developments such as the rise of Islamic fundamentalism in Asia and Africa and the spread of Protestantism in Latin America.[10] Enough said about the relevance of contextualization. There is a tendency to emphasize the situational at the expense of directional factors. I believe that both should be given their due. Let us quickly return to an earlier example: To account for the contrast between positivism and hermeneutical theory, it will not do to simply point to situations in which the one type of theory has more explanatory power than the other, although these contextual considerations should not be neglected. At depth, what is at stake is a clash of views about knowledge and, deeper even, about the place of humans in the cosmos. This conflict, I believe, reaches into deeper realms of human life than sociocultural diversity.

Structures Of-Structures For

I have argued already that worldviews express directions and need to be placed in the proper contexts. Now it is time to consider a third dimension: structures. In this connection structure refers to social institutions with characteristic features, such as families, schools, business enterprises, universities, art institutions, and churches as well as to, in another sense, nations, cities, villages, and clans. In yet another sense, it also refers to distinct spheres of life: economic, social, aesthetic, ethical, and faith spheres.

In the previous section, the contrast between Western and non-Western societies was interpreted only in terms of contextual differences. To take the next step, we have to go beyond distinguishing specific circumstances that show a wide variety of forms and arrangements. The notion of structure refers to differences between the characteristic patterns of traditional and modern societies (for simplicity sake, I equal Western with modern although, of course, this is not entirely correct). One of the hallmarks of a modern society is indeed its differentiated structure. It is only in such a society that, for most people, everyday life characteristically consists of fulfilling various tasks and functions. It is precisely in this respect that a worldview has to fulfill its orientational function. As traditional lifestyles lose their grip, modernization creates massive disorientation. Worldviews (claim to) offer maps showing how to act in accordance with basic principles in very different areas of life; thus helping people to maintain a sense of identity.

What I have just said also applies to the relationship of theory and worldview. Although the distinction that I drew earlier between social worldviews and those in the theoretical realm remains valid; nevertheless, the link with structural differentiation needs to be stressed also with respect to the latter. This is especially true for social theory: From Rousseau on, the great social philosophers have been led by visions of how to make sense of the structural complexities of modern society. Whether one turns to the works of the founders of modern social theory, to Christian thinkers such as Jacques Maritain and Herman Dooyeweerd, or to contemporaries such as Jürgen Habermas, Anthony Giddens, and Charles Taylor, again and again, the overriding concern turns out to be finding coherence and meaning within a structurally differentiated world.

All this goes to show how closely linked worldview and structure are. Yet, it is precisely with respect to structures that a worldview approach reaches its limits. This becomes clear once we take into account a distinction between structures of and structures for, a distinction inspired by a Clifford Geertz essay (1973, 93-95; see also Olthuis 1989, 29); moreover, it runs parallel to the distinction drawn in the Introduction between implications of and implications for).

For structures of, it is appropriate to think of factual regularities. As geographers know only too well, unintended consequences of human action easily gain a (quasi-)normative function: for example, restricting the space for further action, making certain courses of action appear more plausible than others. Regularities created by human action may thus set limits to further action. From the political economists (Adam Smith) of the eighteenth century on, this theme of the unintended consequences has continued to play a pivotal role. However, the question was bound to arise as to whether this is the whole story, whether there is not a deeper kind of normativity, a normativity pertaining to structures that do not emerge simply from the contingencies of human life but function instead as structures for, structures enabling human activities. Philosophers have come up with all kinds of answers. Hegel, for instance, saw the state as a normative structure disclosing human possibilites. Although his idolization of the state (a "heaven on earth") was deeply problematic, yet *au fond* his was a valid insight (reiterated later by Hannah Arendt): The structures of the polis, disclose possibilities for public life that are not possible outside its structures. In the same vein, it is possible to appreciate a search for deep normativity in Durkheim's explorations of a moral reality preceding individual action, or in Habermas's insistence on nonarbitrary (quasi-transcendental) conditions for dialogue. Although their doctrines invite criticisms of many kinds, these thinkers have searched sincerely for what constitutes human society.

Quite naturally, Christians have attributed deep normativity to divine ordinances. In Kuyper, *creational ordinances* is a major theme. Yet, his work also shows the limits of a worldview approach. Conceptual tools are needed to lay bare normative structures, which were only later developed by his successors (Griffioen 1995). Only then did it become possible to overcome the totalism from which Kuyper was unable to free himself. The underlying assumption of Kuyper's two-humanities-two-sciences thesis is that root conflicts (two humanities) are imported without mitigation by worldviews into the system of the sciences (two sciences). One science would be directed toward divine ordinances as normative structures which the other would be bound to ignore in its focus on factual regularities. Instead, a more fruitful approach would be, I believe, to see what consequences root conflicts have for knowing norms (structures for). It would mean testing worldviews to see how rich are the perspectives that they offer. Richness then is more than a matter of contextual diversity, although the boundless physical and human variety that the earth displays should not be downgraded (as geographers will continue to remind us). Richness at a more fundamental level is especially a matter of structures, of spontaneous patterns (structures of), but also of structural norms disclosing possibilities (see for example Janel Curry-Roper's contribution) disclosing possibilities. To discover these structures, painstaking work is needed. Here philosophers and those working in the special disciplines need one another; the critically reflective method (Klapwijk 1994, 166-67, 180, 190) of the one needs to be complementend by the analytical-empirical method of the other.

Vision

Henk Aay's chapter makes clear how difficult, in fact, the transition from worldviews to structures is. His subject is geography education in the Dutch Christian schools in the first half of this century. All ingredients for a succesful transition seemed to be present. Not only did the majority of Dutch children attend Christian schools (as is still the case), but the Protestant schools also were strongly influenced by a neo-Calvinist Christian worldview. Already, at an early stage, it was understood that confessions about the goodness of creation, the distinctiveness of a Christian worldview, the cultural mandate, the notion that life is religion, and the creation order could not but have consequences for geography. The conditions for take-off seemed to be optimal. Yet, as Aay's study of textbooks shows, the results were disappointing. Well-intended efforts notwithstanding, the Christian geographers got stuck in a "blind methodological alley." This road did not lead to a human geography but remained dependent on a sociographic approach. School geography became a branch of comparative sociology and ethnography; a distinct geographic paradigm in education did not develop. Neither did a Christian scholarly and academic geographic tradition emerge. When, in the 1950s, the Free University established its geography department (rather late, as Gerda Hoekveld-Meijer reminds us), there was not a Christian tradition in geography to draw on. Nor did a commitment develop that a Christian university must include human geography in its academic programs. Almost thirty years later, these conditions made it possible for the university to close its geography department (see Gerda Hoekveld-Meijer for details, chap. 5).

Why was it that a Christian mind in geography failed to develop? Was it because of a lack of dedication and vision? In retrospect, it seems that the greatest shortcoming was in what I have called the transition from worldview to structures. To transform the discipline, critical concepts would have been needed, along with new presuppositions, which together might have prompted new approaches in methodology. All this can only happen if the worldview yields structural insights that are strong enough to penetrate into the empirical cycle. Only then do worldviews for not remain disconnected from worldviews of, to reiterate a theme of the Introduction.

To many, I am afraid, all this sounds too utopian to be taken seriously. Let me therefore close by pointing to some real possibilities for renewal. A prerequisite for reform is the presence of a vision that is not a mere perspective but a lived reality. What is needed specifically with respect to Christian scholarship is a community able and willing to nourish a vision for scholarship, or at least willing to support those who have such a vision. I have stressed before that social movements on the whole have not inspired academic renewal. The grand stories of socialism, liberalism, and orthodox Christianity have contributed little to an inner reform of social thought. There is hope as long as communities retain a sense of the unity of life; then scholarship remains a shared concern and Christian schools will provide fertile soil for geography. Teaching geography in primary and secondary schools will then not be just a matter of transferring geographical insights to a nonscholarly context, something that comes at the tail end of other concerns. Rather, the other way around, critical concepts have to take roots in geographic education. This is how I understand what Gerard Hoekveld says in his contribution about education for citizenship. His plea to place ethics before the empirical cycle coincides with my insistence on the need for a worldview community.

There is no guarantee of success. Fellow scholars may prove insensitive to renewal. But the most important test is whether new avenues for scholarship are opened up. The future will show whether this is true of such concepts as externalities (Gerard Hoekveld) or such themes as the primacy of social justice (Gerda Hoekveld-Meijer).

It is encouraging that in as much as the Kuyperian Christian worldview still exerts influence it does so precisely in its impulses for scholarship. In political science, for example, something of an international Christian scholarly community has developed (Griffioen 1995). Why could the same not happen in geography?

Notes

[1] See his remark on mythology in the introduction to the so-called *Grundrisse* (1857): "All mythology overcomes and dominates and shapes the forces of nature in the imagination and by the imagination; it therefore vanishes with the advent of real mastery over them" (Marx 1973, 110).

[2] "Relativism" and "perspectivalism" (or perspectivism) are often treated as synonyms. Alasdair MacIntyre, however, has introduced a subtle distinction: whereas a relativist rejects the possibility of metatraditional criteria to compare rival traditions, a perspectivist denies the possibility of making truth-claims from within a tradition (MacIntyre 1988, 352). See also David Ley's chapter for a cogent critique of perspectivalism.

[3] Recently Richard Mouw has developed such distinctions while commenting on a work by the anthropologist R. Daniel Shaw, *Kandila: Samo Ceremonialism and Interpreted Relationships* (Ann Arbor, 1990), in: Mouw 1996, 7,8.

[4] M. Heidegger, 'Die Zeit des Weltbildes' (1938), in *Holzwege*, Frankfurt: Klostermann. Although in this essay he also uses *Weltanschauung*, Heidegger's prime target is *Weltbild*, in the sense of "the world becoming a picture," instead of a lived reality.

[5] Needham, although himself an agnostic, has to allow for religious reasons. In summarizing the reasons why the Chinese did not develop an idea of universal laws, he mentions as one of the reasons: "[T]he autochthonous ideas of a supreme being, though certainly present from the earliest times, soon lost the qualities of personality and creativity. The development of the concept of precisely formulated abstract laws capable, because of the rationality of an Author of Nature, of being deciphered and re-stated, did not therefore occur. The Chinese worldview depended upon a totally different line of thought . . ." (Needham 1969, 582). An eloquent illustration of the significance of implicit worldviews!

[6] It is true we believe that competition between social movements nowadays has much less of a clash between worldviews than was the case even a few decennia back. This decline, of course, is at the center of Fukuyama's book, *The End of History and the Last Man* (1992).

[7] "The State of Evangelical Christian Scholarship," *The Reformed Journal,* Sept. 1987; quoted in R. D. Henderson 1992, 23. Compare the editorial statement introducing Henderson's essay: "Without question the presuppositional thought of Abraham Kuyper has become extremely influential—some would say dominant—in evangelical scholarship."

[8] Our distinction corresponds with the contrast Alvin Plantinga has drawn between "Perennial Naturalism" and "Creative Anti-Realism" as the two "basic perspectives" or "pictures" of our time (Plantinga 1989-90).

[9] *San Zhi Jing*, i.e., the book of three liners, a once famous primer by Wang Ying Lin (1223-1296) opens as follows: "Men at their birth are naturally good. Their natures are much the same, their habits become widely different." Translation by W. Brooks Brouner and Fung Yuet Mow, in: *Chinese Made Easy* (Leiden: Brill, 1935), 94.

[10] Standard modernization theories have been strongly critized by Samuel Huntington, *The Clash of Civilizations and the Remaking of World Order.* For an interesting discussion of this book, see William McNeill (1997). McNeill accepts Huntington's criticism while holding on to a universalist approach to world history.

About the Contributors

Henk Aay

Henk Aay is professor of geography and environmental studies at Calvin College. His recent research and publications have focussed on the history of geography, ethnic settlement, and ecofiction.

Janel M. Curry-Roper

Janel Curry-Roper is professor of geography and environmental studies at Calvin College. Her major publications are in the areas of land tenure in the Midwest and natural resource policy. Curry-Roper held the Spoelhof Teacher-Scholar chair during the 1996-97 academic year and is presently a scholar of the Calvin Center for Christian Scholarship.

Sander Griffioen

Sander Griffioen is professor of social philosophy at the Free University, Amsterdam. He taught at the Institute for Christian Studies (Toronto) for three years. Besides his professorship at the Free University, he held a special chair for Christian philosophy at the University of Leiden (1979-1990). With Richard J. Mouw he coauthored a study on pluralism.

Gerard A. Hoekveld

Gerard A. Hoekveld is professor of regional geography and geographic education at the State University of Utrecht. Before 1985, he was professor of urban and rural geography at the Free University in Amsterdam, and has published in these fields. Since 1985, he has published regional geographic studies in the context of REGIS, a collaborative research project of these two universities.

Gerda Hoekveld-Meijer

Gerda Hoekveld-Meijer is senior member in geography of education in the Institute of Didactics and Education (IDO). She has published regional geographic studies in the context of REGIS.

David Ley

David Ley is professor of geography at the University of British Columbia and adjunct professor at Regent College. He is also Codirector of the Vancouver Centre for Research on Immigration, where his current work is concerned with immigrant settlement and integration in Canadian cities, including the part played by churches in service provision and in resolving the tension between cultural continuity and cultural integration.

David N. Livingstone

David N. Livingstone is professor of geography at the Queen's University of Belfast and a fellow of the British Academy. He is the author of several books including *The Geographical Tradition* and recently coedited *Human Geography: An Essential Anthology*.

Iain Wallace

Iain Wallace is a professor of geography at Carleton University, Ottawa. He is author of *The Global Economic System* and has recently published in the fields of Canadian environmentalism and social justice in resource-based economies. He is a fellow and governor of the Royal Canadian Geographical Society.

Reference List

Aay, H. 1976. Geography, calling, and curriculum. Part 1. *Christian Educators Journal* 16, no. 1:16-19.

_____. 1977. Geography, calling, and curriculum. Part 2. *Christian Educators Journal* 16, no. 2:11-14.

Adams, P. C. 1995. A reconsideration of personal boundaries in space-time. *Annals of the Association of American Geographers* 85:267-85.

Agnew, J. A. 1989. Sameness and difference: Hartshorne's *The nature of geography* and geography as a real variation. In *Reflections on Richard Hartshorne's "The nature of geography,"* edited by J. N. Entrikin and S. D. Brunn, 121-39. Washington, D.C.: Association of American Geographers.

Agnew, J. A., and S. Corbridge. 1995. *Mastering space: Hegemony, territory, and international political economy.* London: Routledge.

Anderson, K. 1991. *Vancouver's Chinatown: Racial discourse in Canada, 1875-1980.* Montreal: McGill-Queens University Press.

Anderson, W. T. 1990. *Reality isn't what it used to be.* San Francisco: Harper & Row.

Archer, K. 1993. Regions as social organisms: The Lamarckian characteristics of Vidal de la Blanche's regional geography. *Annals of the Association of American Geographers* 38:498-514.

Augustine, *Confessions.* Translated with an introduction by R. S. Pine-Coffin. London: Penguin Books, 1961.

Baldridge, W. 1996. Reclaiming our histories. In *Native and Christian: Indigenous voices on religious identity in the United States and Canada,* edited by J. Treat, 83-92. New York: Routledge.

Barnes, T. J. 1996. *Logics of dislocation: Models, metaphors, and meanings of economic space.* New York: Guilford Press.

Barnes, T. J., and J. Duncan, eds. 1992. *Writing worlds: Discourse, text, and metaphor in the representation of landscape.* London: Routledge.

Barten, A. 1936. Volkenkunde en aardrijkskundeonderricht [Ethnographic and geographic education]. *Tijdschrift voor het Onderwijs in Aardrijkskunde [Journal for Geographic Education]* 14:149-57.

Beauregard, R. 1995. Theorizing the global-local connection. In *World cities in a world-system*, edited by P. Knox and P. Taylor, 232-48. Cambridge: Cambridge University Press.

Bellah, R. N. 1990. The church in tension with a Lockean culture. *New Oxford Review* 57:10-16.

Bellah, R. N., R. Madsen, W. M. Sullivan, A. Swidler, and S. M. Tipton. 1985. *Habits of the heart.* Berkeley and Los Angeles: University of California Press.

Berger, P. L., and T. Luckman. 1966. *The social construction of reality: A treatise in the sociology of knowledge.* New York: Doubleday.

Berlin, I. 1957. *The hedgehog and the fox: An essay on Tolstoy's view of history.* New York: Mentor Books.

Berry, P. 1992. Introduction to *Shadow of spirit: Postmodernism and religion*, edited by P. Berry and A. Wernick, 1-8. London: Routledge.

Berry, P., and A. Wernick, eds.1992. *Shadow of spirit: Postmodernism and religion.* London: Routledge.

Blomley, N. K. 1992. The business of mobility: Geography, liberalism, and the *Charter of Rights. The Canadian Geographer* 36:236-53.

Bolt, J. 1993. *The Christian story and the Christian school.* Grand Rapids: Christian Schools International.

Bonino, J. M. 1975. *Doing theology in a revolutionary situation.* Philadelphia: Fortress.

Botha, M. E. 1988. Objectivity under attack: Rethinking paradigms in social theory, a survey. In *Social science in Christian perspective,* edited by P. A. Marshall and R. E. Vandervennen, 33-62. Lanham, Md: University Press of America.

_____. 1995. The puzzling problem of pluralism. In *Christian philosophy at the close of the twentieth century: Assessment and perspective,* edited by S. Griffioen and B. M. Balk, 159-73. Kampen: Uitgeverij Kok.

Brettler, M. Z. 1995. *The creation of history in ancient Israel.* London: Routledge.

Brewer, Rev. Dr. n.d. [ca.1875]. *Theology in science; or, the testimony of science to the wisdom and goodness of God.* 6th ed. London: Jarrold and Sons.

Brinsmead, R. D. 1978. Man. *Verdict: A Journal of Theology* 1:6-26.

Broek, J. O. M. 1933. De geographische monografie [The geographical monograph]. *Tijdschrift voor het Onderwijs in Aardrijkskunde [Journal for Geographic Education]* 11:31-39.

Brooke, J. H. 1989. Science and the fortunes of natural theology: Some historical perspectives. *Zygon* 24:3-22.

———. 1991. *Science and religion: Some historical perspectives.* Cambridge: Cambridge University Press.

Brubaker, R. 1992. *Citizenship and nationhood in France and Germany.* Cambridge: Harvard University Press.

Byles, J. 1970. Alienated youth. In *The underside of Toronto*, edited by W. Mann, 141-57. Toronto: McClelland and Stewart.

Campbell, C. S. 1994. The second nature of geography: Hartshorne as humanist. *The Professional Geographer* 46:411-17.

Campbell, J. 1990. Personhood and the land. *Agriculture and Human Values* 7:39-43.

Centraal Bureau voor de Statistiek [Central Bureau for Statistics]. 1940. *Statistisch zakboek, 1940 [Statistical notebook, 1940]*. The Hague: Albani.

———. 1970. *Statistisch zakboek, 1970 [Statistical notebook, 1970]*. The Hague: Staatsuitgeverij.

Chorley, R. J., R. P. Beckinsale, and A. J. Dunn, eds. 1973. *The history of the study of landforms or the development of geomorphology.* Vol. 2, *The life and work of William Morris Davis.* London: Methuen.

Christie, N. J. 1994. Environment and race: Geography's search for a Darwinian synthesis. In *Darwin's laboratory: Evolutionary theory and natural history in the Pacific*, edited by R. MacLeod and P. F. Rehbock, 426-73. Honolulu: University of Hawaii Press.

Clark, J. C. D. 1985. *English society, 1688-1832: Ideology, social structure, and political practice during the ancient régime.* Cambridge: Cambridge University Press.

_____. 1989. England's ancient régime as a confessional state. *Albion* 21:450-74.

Clemenger, B. 1994. "Von Mises" economic reductionism. In *Political theory and Christian vision: Essays in memory of Bernard Zylstra,* edited by J. Chaplin and P. Marshall, 231-46. Lanham, Md.: University Press of America.

Clifford, J. 1986. Introduction: Partial truths. In *Writing culture,* edited by J. Clifford and G. Marcus, 1-26. Berkeley and Los Angeles: University of California Press.

Clines, D. J. A. 1978. *The theme of the Pentateuch.* JSOT supp. 10. Sheffield: Sheffield Academic Press.

Clouser, R. A. 1983. Religious language: A new look at an old problem. In *Rationality in the Calvinian tradition,* edited by H. Hart, J. Van Der Hoeven, and N. P. Wolterstorff, 385-407. Lanham, Md.: University Press of America.

_____. 1991. *The myth of religious neutrality: An essay on the hidden role of religious belief in theories.* Notre Dame: University of Notre Dame Press.

Cohen, L. 1924. Reviews of *De werelddeelen: Beknopt leerboek der aardrijkskunde [World regions: Concise textbook for geography]* and *Europa: Beknopt leerboek der aardrijkskunde [Europe: Concise textbook for geography],* by L. Romkes. *Tijdschrift voor het Onderwijs in Aardrijkskunde [Journal for Geographic Education]* 2:31-36.

_____. 1931a. Naar de verzoening? [Towards reconciliation?] *Tijdschrift voor het Onderwijs in Aardrijkskunde [Journal for Geographic Education]* 9:121-25.

_____. 1931b. Sociografie, sociologie, en sociale aardrijkskunde [Sociography, sociology, and social geography]. *Tijdschrift voor het Onderwijs in Aardrijkskunde [Journal for Geographic Education]* 9:169-75.

Corbridge, S., R. Martin, and N. Thrift, eds. 1994. *Money, power, and space.* Oxford: Blackwell.

Cosgrove, D. 1990. Environmental thought and action: Premodern and postmodern. *Transactions of the Institute of British Geographers* n.s. 15:344-58.

Cox, K. R. 1972. *Man, location, and behavior: An introduction to human geography*. New York: John Wiley and Sons.

Cremer A. F., and J. Ligthart. 1909. *Op de fiets door Nederland* [*On the bike through the Netherlands*]. Groningen: Wolters.

Daly, H. E., and J. B. Cobb Jr. 1989. *For the common good: Redirecting the economy toward community, the environment, and a sustainable future.* Boston: Beacon Press.

Daniels, S. 1992. Place and the geographical imagination. *Geography* 77:310-22.

Davie, G. E. 1994. Victor Cousin and the Scottish philosophers. In *A passion for ideas: Essays on the Scottish Enlightenment.* Vol. 2. Edinburgh: Polygon.

Davies, G. L. 1969. *The earth in decay: A history of British geomorphology 1578-1878.* London: MacDonald.

Davies, P. 1984. *God and the new physics.* Harmondsworth: Pelican.

Davis, W. M. 1934. The faith of reverent science. In *The history of the study of landforms or the development of geomorphology.* Vol. 2, *The life and work of William Morris Davis*, edited by R. J. Chorley, R. P. Beckinsale, and A. J. Dunn, 1973, 759-91. London: Methuen. Originally published in *Scientific Monthly* 38 (1934): 395-421.

Day, G., and J. Murdoch. 1993. Locality and community: Coming to terms with place. *The Sociological Review* 41:82-111.

De Graaff, A. H., J. Olthuis, and A. Tuininga. 1980. *Japan: A way of life.* Toronto: Joy in Learning Curriculum Development and Training Center.

_____. 1981. *Kenya: A way of life.* Toronto: Joy in Learning Curriculum Development and Training Center.

De Jong, Gerben. 1955. *Het karakter van de geografische totaliteit* [*The nature of the geographic whole*] Groningen: Wolters.

_____. 1958. The nature of human geography in the light of the ordinances of creation. *Free University Quarterly* 5:97-119.

_____. 1962. *Chorological differentiation as the fundamental principle of geography: An inquiry into the chorological conception of geography.* Groningen: Wolters.

Derham, W. 1727. *Physico-theology: or, a demonstration of the* being *and* attributes *of God, from his works of creation. Being the substance of sixteen sermons preached in St. Mary-le-Bow-Church, London; At the Honorable Mr. Boyle's Lectures, in the years 1711, and 1712.* 8th ed. London.

Deurloo, K. A. 1990. Narrative geography in the Abraham cycle. *Old Testament Series* 26:48-62.

Deutsche, R. 1991. Boys' town. *Society and Space* 9:5-30.

Dick, T. 1825. *The Christian philosopher; or, the connection of science and philosophy with religion.* 3d ed. Glasgow: Chalmers and Collins.

Dorsey, D. A. 1991. *The roads and highways of ancient Israel.* Baltimore: Johns Hopkins University Press.

Duncan, J., and D. Ley. 1993. Introduction: Representing the place of culture. In *Place/culture/representation*, edited by J. Duncan and D. Ley, 1-21. London: Routledge.

Duncan, S., and M. Savage. 1989. Space, scale, and locality. *Antipode* 21:179-206.

Eagleton, T. 1990. *The ideology of the aesthetic.* Oxford: Blackwell.

Edelman, D. V. 1991. *The fabric of history: Text, artifact, and Israel's past.* JSOT supp. 127. Sheffield: Sheffield Academic Press.

Engelken, K. 1990. Kanaan als nicht-territorialer Terminus. *Biblische Notizen* 52:47-63.

Entrikin, J. N. 1984. Carl O. Sauer: Philosopher in spite of himself. *Geographical Review* 74:387-408.

_____. 1991. *The betweenness of place: Towards a geography of modernity.* Baltimore: Johns Hopkins University Press.

Fahrenfort, J. J. 1926. Middelbare wetenschap [Secondary school knowledge]. *Tijdschrift voor het Onderwijs in Aardrijkskunde* [*Journal for Geographic Education*] 4:123-31.

Fernhout, H. 1990. The metaphor of "development" in moral education. In *Norm and context in the social sciences*, edited by S. Griffioen and J. Verhoogt, 75-108. Lanham, Md.: University Press of America.

Finkel, A. 1984. Gerechtigkeit: Judentum. In *Theologische Realencyklopedie.* Band 12, 414-20. Berlin: Walter de Gruyter.

Flint, R. 1879. *Anti-theistic theories.* Edinburgh: Blackwood and Sons.

_____. 1883. *Theism, being the Baird lecture for 1876.* Edinburgh: Blackwood and Sons.

Fraser, N. 1991. Rethinking the public sphere: A contribution to the critique of actually existing democracy. In *Habermas and the public sphere,* edited by C. Calhoun. Cambridge, Mass.: MIT Press.

Friedland, W. H. 1994. The new globalization: The case of fresh produce. In *From Columbus to ConAgra: The globalization of agriculture and food,* edited by A. Bonanno et al., 210-31. Lawrence, Kans.: University of Kansas Press.

Fukuyama, F. 1992. *The end of history and the last man.* New York: Free Press.

Gadamer, H.-G. 1986. *Truth and method.* New York: Crossroads.

Garcia, S. 1996. Cities and citizenship. In *International Journal of Urban and Regional Research* 20, no.1:7-22.

Garsiel, M. 1987. *Biblical names: A literary study of Midrashic derivations and puns.* Ramat Gan: Bar-Ilan University Press.

_____. 1991. Puns upon names as a literary device in 1 Kings 1-2. *Biblica* 72:379-86.

Geertz, C. 1973. *The interpretation of cultures.* New York: Basic Books.

Gereformeerd Schoolverband. 1955. *Aanvullende gegevens bij het onderwijs in de aardrijkskunde aan Christelijke scholen* [*Supplementary materials for geographic education in Christian schools*]. Kampen: Kok.

Gilhuis, T. M. 1974. *Memorietafel van het Christelijk onderwijs: De geschiedenis van de Christelijk schoolstrijd* [*Significant moments in Christian education: The history of the Christian school question*]. Kampen: Kok.

Glacken, C. J. 1967. *Traces on the Rhodian shore: Nature and culture in western thought from ancient times to the end of the eighteenth century.* Berkeley and Los Angeles: University of California Press.

Gottmann, J. 1952. *La politique des États et leur Géographie*. Paris: Librairie Armand Colin.

Goudzwaard, R. 1992. Creation management: The economics of earth stewardship. *Firmament* (winter): 4-5, 21-23.

Goudzwaard, R., and H. De Lange. 1995. *Beyond poverty and affluence: Toward an economy of care*. Grand Rapids: Eerdmans.

Granovetter, M. 1985. Economic action and social structure: The problem of embeddedness. *American Journal of Sociology* 91:481-510.

Gray, C. M. 1991. *With Liberty and justice for whom? The recent evangelical debate over capitalism*. Grand Rapids: Eerdmans.

Gregory, D. 1994. *Geographical imaginations*. Oxford: Blackwell.

Gregory, D., R. Martin, and G. Smith, eds. 1994. *Human geography: Society, space, and social science*. Minneapolis: University of Minnesota Press.

Griffioen, S. 1986. De betekenis van Dooyeweerd's ontwikkelingsidee [The meaning of Dooyeweerd's concept of development]. *Philosophia Reformata* 51, nos. 1-2.

_____. 1994. Is a pluralist ethos possible? *Philosophia Reformata* 59:11-25.

_____. 1995a. Kleine typologie van pluraliteit [A small typology of plurality]. In *Pluralisme, cultuur filosofische beschouwingen* [Pluralism, cultural philosophical considerations], edited by T. DeBoer and S. Griffioen, 204-26. Meppel: Boom.

_____. 1995b. The relevance of Dooyeweerd's theory of social institutions. In *Christian philosophy at the close of the twentieth century*, edited by S. Griffioen and B. M. Balk, 139-58. Kampen: Kok.

Griffioen, S., and B. M. Balk, eds. 1995. *Christian philosophy at the close of the twentieth century*. Kampen: Kok.

Griffioen, S., and J. Verhoogt. 1990. Introduction: Normativity and contextuality in the social sciences. In *Norm and context in the social sciences*, edited by S. Griffioen and J. Verhoogt, 9-22. Lanham, Md.: University Press of America.

Gundlach, B. J. 1995. The evolution question at Princeton, 1845-1929, Ph.D. diss., University of Rochester, New York.

Gunson, N. 1994. British missionaries and their contribution to science in the Pacific Islands. In *Darwin's laboratory: Evolutionary theory and natural history in the Pacific*, edited by R. MacLeod and P. F. Rehbock, 283-316. Honolulu: University of Hawaii Press.

Gurvitch, G. 1963. Social structure and multiplicity of times. In *Sociological theory, values, and sociocultural change*, edited by E. A. Tiriakyan, 171-84. New York: Free Press of Glencoe.

Guyot, A. 1884. *Creation; or, the biblical cosmogony in the light of modern science*. New York: Charles Scribner's Sons.

_____. 1897. *The earth and man: Lectures on comparative physical geography in its relation to the history of mankind*. New York: Charles Scribner's Sons.

Habermas, J. 1982. *Theorie des kommunikativen Handelns*. Vol. 2. Frankfurt: Suhrkamp.

_____. 1992. Citizenship and national identity: Some reflections on the future of Europe. *Praxis International* 12, no.1:1-19.

Hall, D. J. 1988. The spirituality of the covenant: Imaging God, stewarding earth. *Perspectives* (December): 11-14.

_____. 1991. *Thinking the faith: Christian theology in a North American context*. Minneapolis: Fortress.

Hanrath, J. J. 1926. Middelbare wetenschap of wetenschap? [Secondary school knowledge or science?]. *Tijdschrift voor het Onderwijs in Aardrijkskunde* [*Journal for Geographic Education*] 4:160-64.

Haraway, D. 1988. Situated knowledges: The science question in feminism and the privilege of partial perspective. *Feminist Studies* 14:575-99.

_____. 1996. Situated knowledges: The science question in feminism and the privilege of partial perspective. In *Human geography: An essential anthology*, edited by J. Agnew, D. Livingstone, and A. Rogers, 108-28. Oxford: Blackwell.

Harrison, B. 1992. Industrial districts: Old wine in new bottles? *Regional Studies* 26:469-83.

Harrison, R. T., and D. Livingstone. 1980. Philosophy and problems in human geography: A presuppositional approach. *Area* 12, no. 1:25-31.

Hart, H. 1984. *Understanding our world: An integral ontology*. Lanham, Md.: University Press of America.

Hartshorne, R. 1939. *The nature of geography*. Lancaster, Pa.: Association of American Geographers.

Harvey, D. 1973. *Social justice and the city*. London: Edward Arnold.

_____. 1978. The redistribution of real income in an urban system. In *Fundamentals of Human Geography: A Reader*, edited by J. Blunden and P. Haggett, 277-84. London: Harper & Row.

_____. 1989. *The condition of postmodernity*. Oxford: Blackwell.

_____. 1992. Social justice, postmodernism, and the city. *International Journal of Urban and Regional Research* 16:588-601.

_____. 1993. Class relations, social justice, and the politics of difference. In *Principled positions: Postmodernism and the rediscovery of value*, edited by J. Squires, 85-120. London: Lawrence and Wishart.

Hastings, A. 1996. Religion, ethnicity, and national identity. Series presented for the Wiles Lectures at Queens University of Belfast.

Heater, D. 1990. *Citizenship: The civic ideal in world history, politics, and education*. London: Longman.

Henderson, R. D. 1992. How Abraham Kuyper became a Kuyperian. *Christian Scholar's Review* 22 (September): 22-35.

Hoekveld, G. A. 1990. Regional geography must adjust to new realities. In *Regional geography, current developments, and future prospects*, edited by R. J. Johnston, J. Hauer, and G. A. Hoekveld, 11-31. London: Routledge.

Hoekveld-Meijer, G. 1996. *Esau—Salvation in Disguise: Genesis 36, a hidden polemic between our teacher and the prophets about Edom's role in postexilic Israel*. Kampen: Kok Pharos.

Homer-Dixon, T. F. 1993. Environmental scarcity and violent conflict. *Scientific American* 268, no. 2:38-45.

Hoogwerf, J. 1918. Aardrijkskunde op de Christelijke school [Geography in the Christian school]. *Paedagogisch Tijdschrift voor het Christelijk Onderwijs* [*Pedagogical Journal for Christian Education*] 11:212-24.

Hoogwerf J., and D. J. Baarslag. 1929a. *Aardrijkskundig leesboek voor het Christelijk onderwijs* [*Geography reader for Christian education*]. 8 vols. Groningen: Noordhoff.

_____. 1929b. *Bij de kaart: Eenvoudige aardrijkskunde voor het Christelijk Onderwijs* [*Close to the map: Simple geography for Christian education*]. 3 vols. Groningen: Noordhoff.

hooks, b. 1992. Representing whiteness. In *Cultural studies,* edited by L. Grossberg, C. Nelson, and P. Treichler, 338-46. New York: Routledge.

Houtman, C. 1982. *Wereld en tegenwereld: Mens en milieu in de bijbel / mens en milieu en de bijbel* [*World and antiworld: Humanity and nature in the Bible/humanity and nature and the Bible*]. Baarn: Ten Have.

_____. [1986] 1993. *Exodus 1.* Kampen: Kok.

Howland, D. R. 1996. *Borders of Chinese civilization: Geography and history at empire's end.* Durham: Duke University Press.

Hummon, D. M. 1990. *Common places: Community ideology and identity in American culture.* Albany: State University of New York Press.

Jacob, M. 1976. *The Newtonians and the English revolution 1689-1720.* Ithaca: Cornell University Press.

Jacobs, J. 1996. *Edge of empire: Postcolonialism and the city.* London: Routledge.

Janse, A. 1935. *Het eigen karakter der Christelijke school* [*The unique character of the Christian school*]. Kampen: Kok.

Jeffrey, D. L. 1987. *English spirituality in the age of Wesley.* Grand Rapids: Eerdmans.

Johnston, R. J., P. J. Taylor, and M. J. Watts, eds. 1995. *Geographies of global change: Remapping the world in the late twentieth century.* Oxford: Blackwell.

Keck, L. E. 1996. Rethinking New Testament ethics. *Journal of Biblical Literature* 115:3-16.

Keller, E. F. 1985. *Reflections on gender and science.* New Haven: Yale University Press.

Klapwijk, J. 1994. Pluralism of norms and values: On the claim and reception of the universal. *Philosophia Reformata* 59:158-92.

Kline, M. G. 1972. *The structure of biblical authority*. Grand Rapids: Eerdmans.

Kloppenburg, J., Jr. 1991. Social theory and the de/reconstruction of agricultural science: Local knowledge for an alternative agriculture. *Rural Sociology* 56:519-48.

Knox, P., and J. Agnew. 1994. *The geography of the world economy*. 2d ed. London: Arnold.

Kothari, S., and P. Parajuli. 1993. No nature without social justice: A plea for cultural and ecological pluralism in India. In *Global ecology: A new arena of political conflict*, edited by W. Sachs, 224-41. London: Zed.

Kremers, H. 1962. Reich Gottes im Alten Testament und Spätjudentum. In *Evangelischer Kirchenlexicon*. Band 3, 553-55. Gössinger: Van den Hoeck & Ruprecht.

Kruijt, J. P. 1934. Het object van de menselijke aardrijkskunde [The object of human geography]. *Tijdschrift voor het Onderwijs in Aardrijkskunde* [*Journal for Geographic Education*] 12:73-85.

Kuperus, G. 1926. Wetenschappelijke aardrijkskunde [Scientific geography]. *Tijdschrift voor het Onderwijs in Aardrijkskunde* [*Journal for Geographic Education*] 4:131-35.

_____. 1934. Sociografie en geografie [Sociography and geography]. *Tijdschrift voor het Onderwijs in Aardrijkskunde* [*Journal for Geographic Education*] 12:157-62.

Kuyvenhoven, A. 1974. *Partnership: A study of the covenant*. Grand Rapids: Christian Reformed Church Publications.

Kymlicka, W., and W. Norman 1994. Return of the citizen: A survey of recent work on citizenship theory. *Ethics* 104:352-81.

Langedijk, D. 1953. *De geschiedenis van het Protestants-Christelijk onderwijs* [*The history of Protestant Christian education*]. Delft: Van Keulen.

Lankamp, H. n.d. *Aardrijkskunde op de Christelijke school* [*Geography in the Christian school*]. 2 vols. Groningen: Noordhoff.

Larson, J. 1986. Not without a plan: Geography and natural history in the late eighteenth century. *Journal of the History of Biology* 19:447-88.

Layendekker, L. 1986. *Brengt de vooruitgang ons verder? Kanttekeningen bij een wijd en verbreid geloof* [*Does progress bring us further? Short comments on a broad and popular belief*]. Baarn: Ten Have.

Levy, S. J. 1995. Judaism, population, and the environment. In *Population, consumption, and the environment: Religious and secular responses*, edited by H. Coward, 73-107. Albany: State University of New York Press.

Lewthwaite, G. R. 1971. A geographer addresses the question. In *Why I am still a Christian*, edited by E. M. Blaiklock, 110-23. Grand Rapids: Zondervan.

_____. 1973. Geography. In *Christ and the modern mind*, edited by R. W. Smith, 175-84. Downers Grove, Ill.: InterVarsity Press.

Ley, David. 1983. *A social geography of the city*. New York: Harper & Row.

_____. 1996. *The new middle class and the remaking of the inner city*. Oxford: Oxford University Press.

Livingstone, D. 1857. *Missionary travels and researches in South Africa; including a sketch of sixteen years' residence in the interior of Africa*. London: John Murray.

Livingstone, D. N. 1984. Natural theology and neo-Lamarckism: The changing context of nineteenth century geography in the United States and Great Britain. *Annals of the Association of American Geographers* 74:9-28.

_____. 1991. The moral discourse of climate: Historical considerations on race, place, and virtue. *Journal of Historical Geography* 17:413-34.

_____. 1992. *The geographical tradition: Episodes in the history of a contested enterprise*. Oxford: Blackwell.

_____. 1995. The polity of nature: Representation, virtue, strategy. *Ecumene* 2:353-77.

_____. 1996. High tea at the cyclotron. *Books and Culture* 2, no. 1:22-23.

_____. In press. Geography and natural theology. In *History of the geosciences: An encyclopedia*, edited by G. Good. New York: Garland Publishing.

Lowenthal, D. 1961. Geography, experience, and imagination: Toward a geographical epistemology. *Annals of the Association of American Geographers* 51:241-60.

Lührmann, D. 1984. Gerechtigkeit: Neues Testament. In *Theologische Realencyklopedie*. Band 12, 414-20. Berlin: Walter de Gruyter.

Lukermann, F. 1989. *The nature of geography*: post hoc, ergo propter hoc? In *Reflections on Richard Hartshorne's "The nature of geography,"* edited by J. N. Entrikin and S. D. Brunn, 53-68. Washington, D.C.: Association of American Geographers.

Lundin, R. 1993. *The culture of interpretation: Christian faith and the postmodern world*. Grand Rapids: Eerdmans. Quoting S. Nietzche, *The portable Nietzche*. Edited by Walter Kaufmann. New York: Viking, 1954, p. 46-47.

Lyotard, J-F. 1984. *The postmodern condition*. Minneapolis: University of Minnesota Press.

McIntire, C. T. 1985. Dooyeweerd's philosophy of history. In *The Legacy of Herman Dooyeweerd*, edited by C. T. McIntire. Lanham, Md.: University Press of America.

MacIntyre, A. 1988. *Whose justice? Which rationality?* Notre Dame: University of Notre Dame Press.

_____. 1990. *Three rival versions of moral enquiry: Encyclopedia, genealogy, and tradition*. London: Duckworth.

McNeill, W. H. 1995. *Keeping together in time: Dance and drill in human history*. Cambridge: Harvard University Press.

_____. 1997. Decline of the west? *New York Review of Books*, 9 January, 18-22.

Macpherson, H. 1951. Thomas Dick: "The Christian philosopher." *Records of the Scottish Church History Society* 11:41-62.

McPherson, T. 1972. *The argument from design*. London: Macmillan.

Maffesoli, M. 1996. *The time of the tribes*. London: Sage.

Mannheim, K. 1952. *Essays on the sociology of knowledge*. London: Routledge & Kegan Paul.

Marshall, P. 1986. *Thine is the kingdom*. Grand Rapids: Eerdmans.

_____. 1990. Justice and rights: Ideology and human rights theories. In *Norm and context in the social sciences*, edited by S. Griffioen and J. Verhoogt, 129-58. Lanham, Md.: University Press of America.

Marshall, T. H. 1950. *Citizenship and social class and other essays*. Cambridge: Cambridge University Press.

Martin, R. 1994. Economic theory and human geography. In *Human Geography*, edited by D. Gregory, R. Martin, and G. Smith, 21-53. Minneapolis: University of Minnesota Press.

Marx, K. 1973. *Grundrisse: Foundations of the critique of political economy*. Translated by Martin Nicolaus. New York: Random House.

Massey, D. 1991a. Flexible sexism. *Society and Space* 9:31-57.

_____. 1991b. The political place of locality studies. *Environment and Planning A* 23:267-81.

_____. 1993. Questions of locality. *Geography* 78:142-49.

Matless, D. 1991. Nature, the modern, and the mystic: Tales from early twentieth century geography. *Transactions of the Institute of British Geographers* n.s. 16:272-86.

Maury, M. F. [1855] 1874. *The physical geography of the sea*. London: T. Nelson.

Mayhew, R. 1996. Landscape, religion, and knowledge in eighteenth-century England. *Ecumene* 3:454-71

Mehmet, O. 1995. *Westernizing the third world: The eurocentricity of economic development theories*. London: Routledge.

Messelink, J. 1994. Vijftig jaar Gereformeerd onderwijs: Achtergrond, motieven, en opgaven [Fifty years of Gereformeerd education: Background, motives, and tasks] in *1944 en vervolgens: Tien maal over vijftig jaar Vrijmaking* [*1944 and then: Ten times about fifty years Vrijmaking*], edited by G. Harinck, M. te Velde, 29-41. Barneveld: De Vuurbaak.

Middleton, J. R., and B. Walsh. 1995. *Truth is stranger than it used to be: Biblical faith in a postmodern age*. Downers Grove, Ill.: InterVarsity Press.

Milbank, J.1990. *Theology and social theory: Beyond secular reason.* Oxford: Blackwell.

Mill, H. R. 1929. Geography. In *Encyclopaedia Britannica*, 14th ed. Vol. 10, London: The Encyclopedia Britannica Company, q.v.

Miller, B. 1992. Collective action and rational choice: Place, community, and the limits to individual self-interest. *Economic Geography* 68:22-42.

Miller, J. 1993. Missions, social change, and resistance to authority: Notes toward an understanding of the relative autonomy of religion. *Journal for the Scientific Study of Religion* 32:29-50.

Mouw, R. J. 1996. Christian scholarship: The difference a worldview makes. Address at the 6th annual national conference of the Lilly Fellows Program, 18-20 October. Valparaiso University, Valapraiso, Ind.

Mouw, R. J., and S. Griffioen. 1993. *Pluralisms and horizons.* Grand Rapids: Eerdmans.

Mulder, H. 1988. *Biografisch lexicon voor de geschiedenis van het Nederlandse protestantisme* [*Biographical dictionary for the history of Dutch Protestantism*] s. v. Van Deursen, Arie. Vol. 3. Kampen: Kok.

National Geographic Research and Exploration 1994. *Geography for Life, National Geography Standards 1994.* Washington, D.C.: National Geographic.

Needham, J. 1969. *Science and Civilization in China.* Vol. 2, *History of Scientific Thought.* Cambridge: Cambridge University Press.

Newbigin, L. 1986. *Foolishness to the Greeks: The gospel and western culture.* Grand Rapids: Eerdmans.

_____. 1989. *The gospel in a pluralist society.* Grand Rapids: Eerdmans.

Noble, D. F. 1995. *Progress without people: New technology, unemployment, and the message of resistance.* Toronto: Between the Lines.

Nussbaum, M. 1995. Patriotisme en kosmopolitanisme [Patriotism and cosmopolitanism], Dutch translation of a chapter forthcoming in *Poetic justice*, Boston: Beacon Press. Originally published in *Nexus* 11:18-30.

Olthuis, J. H. 1989. On worldviews. In *Stained glass*, edited by P. A. Marshall, S. Griffioen, and R. J. Mouw, 26-40. Lanham, Md.: University Press of America.

O'Neill, J. 1994. *The missing child in liberal theory: Towards a covenant theory of family, community, welfare, and the civic state.* Toronto: University of Toronto Press.

Pasewark, K. 1993. *A theology of power: Being beyond domination.* Minneapolis: Fortress Press.

Peacocke, A. 1985. Biological evolution and Christian theology: Yesterday and today. In *Darwinism and divinity,* edited by J. R. Durant, 101-30. Oxford: Blackwell.

Peet, R. 1985. The social origins of environmental determinism. *Annals of the Association of American Geographers* 75:309-33.

_____, ed. 1977. *Radical geography: Alternative viewpoints on contemporary social issues.* Chicago: Maaroufa Press.

Perkins, P. 1992. Ethics: New Testament. In *The Anchor Bible Dictionary.* Vol. 1, edited by D. N. Freedman, 652-65. New York: Doubleday.

Perrin, N. 1963. *The kingdom of God in the teaching of Jesus.* London: SCM Press.

Plantinga, A. 1982. The reformed objection to natural theology. *Christian Scholar's Review* 11:187-98.

_____. 1983. Reason and belief in God. In *Faith and rationality: Reason and belief in God,* edited by A. Plantinga and N. P. Wolterstorff, 16-93. Notre Dame: University of Notre Dame Press.

_____. 1984. Advice to Christian philosophers. *Faith and Philosophy* 1:253-71.

_____. 1989-1990. *The twin pillars of Christian scholarship.* Paper read at the Stob Lectures, Calvin College and Seminary, Grand Rapids, Mich.

Polanyi, M. 1958. *Personal knowledge.* Chicago: University of Chicago Press.

Polkinghorne, J. 1986. *One world: The interaction of science and theology.* London: SPCK.

Price, M., and M. Lewis. 1993. The reinvention of cultural geography. *Annals of the Association of American Geographers* 83:1-17.

Rawls, J. 1971. *A theory of justice.* Cambridge: Harvard University Press.

Reid, T. R. 1995. Confucius says: Go east, young man. *Guardian Weekly* 31 (December): 12.

Rich, A. 1980. Sozialethische Kriterien und Maximen humaner Gesellschaftsgestaltung. In *Herausg. Christliche Wirtschaftsethik vor neuen Aufgaben,* edited by T. Strom, 17-28. Zürich: Theologischer Verlag.

Romkes, L. 1924a. *De werelddeelen: Beknopt leerboek der aardrijkskunde* [*World regions: Concise textbook for geography*]. Zwolle: Tjeenk Willink.

_____. 1924b. *Europa: Beknopt leerboek der aardrijkskunde* [*Europe: Concise textbook for geography*]. Zwolle: Tjeenk Willink.

_____. 1927. *Beknopt Leerboek der Aardrijkskunde van Nederland* [*Concise textbook for the geography of the Netherlands*]. Zwolle: Tjeenk Willink.

_____. 1929-30. Is een aardrijkskunde-leerboek voor onze inrichtingen, Christelijk middelbaar onderwijs mogelijk? [Is a geography textbook for our Christian secondary education institutions possible?] *Christelijk Middelbaar Onderwijs* [*Christian Secondary Education*] 10:421-25 .

Rorty, R. 1991. *Objectivity, relativism, and truth.* Cambridge: Cambridge University Press.

Rose, G. 1993. *Feminism and geography: The limits of geographical knowledge.* Minneapolis: University of Minnesota Press.

Rupke, N. In press. Introduction to *Kosmos* by Alexander von Humbold. Baltimore: Johns Hopkins University Press.

Russell, C. A. 1989. The conflict metaphor and its social origins. *Science and Christian Belief* 1:3-26.

Ryan, W. F. 1995. *Culture, spirituality, and economic development: Opening a dialogue.* Ottawa: International Development Research Centre.

Sack, R. D. 1989. The nature, in light of the present. In *Reflections on Richard Hartshorne's "The nature of geography,"* edited by J. N. Entrikin and S. D. Brunn, 141-62. Washington, D.C.: Association of American Geographers.

_____. 1993. The power of place and space. *Geographical Review* 83:326-29.

Sanneh, L. 1987. Christian missions and the western guilt complex. *The Christian Century,* 8 April, 330-35.

_____. 1993. *Encountering the west, Christianity and the global cultural process: The Africa dimension.* London: Marshall Pickering.

Sauer, C. O. 1963. The morphology of landscape. In *Land and life,* edited by J. Leighly, 315-50. Berkeley and Los Angeles: University of California Press.

Sayer, A. 1991. Behind the locality debate: Deconstructing geography's dualisms. *Environment and Planning* A 23:283-308.

_____. 1993. Postmodernist thought in geography: A realist view. *Antipode* 25:320-44.

Schaefer, F. K. 1953. Exceptionalism in geography: A methodological examination. *Annals of the Association of American Geographers* 43:226-49.

Schama, S. 1987. *The embarrassment of riches: An interpretation of Dutch culture in the Golden Age.* New York: Knopf.

Schenk. W. 1983. *Evangelium - Evangelien - Evangeliologie. Ein hermeneutisches Manifest.* München: Chr. Kaiser Verlag.

Schnackenburg, R. 1986. *Die sittliche Botschaft des Neuen Testaments.* Band 1, *Von Jesus zur Urkirche.* Freiburg: Herder.

_____. 1988. *Diesittlicke Botschaft des Neuen Testments.* Band 2, *Die Urchristlichen Verkündiger.* Freiburg: Herder.

Schrage, W. 1982. Ethik, Neues Testament. In *Theologische Realencyklopädie,* Band 10, 425-62. Berlin: Walter de Gruyter.

Selznick, P. 1992. *The moral commonwealth: Social theory and the promise of community.* Berkely and Los Angeles: University of California Press.

Shapin, S. 1981. Of gods and kings: Natural philosophy and politics in the Leibniz-Clarke disputes. *Isis* 72:187-215

_____. 1994. *A social history of truth: Civility and science in seventeenth century England.* Chicago: University of Chicago Press.

Shin, K. W. 1994. *A hermeneutic utopia: H.-G. Gadamer's philosophy of culture.* Toronto: Tea for Two.

Sire, J. W. 1988. *The universe next door: A basic worldview catalog.* 2d ed. Downers Grove, Ill.: InterVarsity Press.

Skillen, J. W. 1994. Toward a comprehensive science of politics. In *Political theory and Christian vision: Essays in memory of Bernard Zylstra,* edited by J. Chaplin and P. Marshall, 57-80. Lanham, Md.: University Press of America.

Skinner, Q. 1985. *The return of grand theory in the human sciences.* Cambridge: Cambridge University Press.

Smelser, N. 1995. Economic rationality as a religious system. In *Rethinking materialism,* edited by R. Wuthnow, 73-92. Grand Rapids: Eerdmans.

Smend, R. 1982. Ethik, Altes Testament. In *Theologische Realencyklopädie.* Band 10, 423-35. Berlin: Walter de Gruyter.

Smilde, H. 1963-64. In *Jaarboek van de maatschappij van Nederlandse letterkunde* [*Yearbook of the society of Dutch literature*], s.v. A. Van Deursen, 55-62 (including a partial bibliography). Leiden: Brill.

Smith, D. M. 1979. *Where the grass is greener: Living in an unequal world.* New York: Penguin Books.

Smith, M. P. 1995. The disappearance of world cities and the globalization of local politics. In *World cities in a world-system,* edited by P. Knox and P. Taylor, 249-66. Cambridge: Cambridge University Press.

Smith, N. 1989. Geography as museum: Private history and conservative idealism in *The nature of geography.* In *Reflections on Richard Hartshorne's "The nature of geography,"* edited by J. N. Entrikin and S. D. Brunn, 91-120. Washington, D.C.: Association of American Geographers.

Smith, S. Y. 1989. Society, space, and citizenship: A human geography for the "new times"? *Transactions of the Institute of British Geographers* 14:144-56.

_____. 1994. *Geography and social justice.* Oxford: Blackwell.

_____. 1995. Moral teaching in geography. *Journal of Higher Education* 66:271-83.

Spate, O. H. K. 1960. Quantity and quality in geography. *Annals of the Association of American Geographers* 50:377-94.

Stallings, B. 1995. The new international context of development. In *Global change, regional response: The new international context of development,* edited by B. Stallings, 349-87. Cambridge: Cambridge University Press.

Stauffer, E. 1959. *Die Botschaft Jesu, damals und heute.* Bern: Francke Verlag.

Steinmetz, S. R. 1930. De betekenis der geographische vakken als wetenschappen [The meaning of the geographical disciplines as sciences]. *Tijdschrift voor het Onderwijs in Aardrijkskunde [Journal for Geographic Education]* 8:169-73.

Stenhouse, J. 1994. The Darwinian enlightenment and New Zealand politics. In *Darwin's laboratory: Evolutionary theory and natural history in the Pacific,* edited by R. MacLeod and P. F. Rehbock. Honolulu: University of Hawaii Press.

Stilma, L. C. 1987. De school met de bijbel in historisch-pedagogisch perspectief [The school with the Bible in historical-pedagogical perspective]. Ph.D. diss., University of Amsterdam.

_____. 1995. *Van kloosterklas tot basisschool: Een historisch verzicht van opvoeding en onderwijs in Nederland [From monastery class to elementary school: A historical overview of nurture and education in the Netherlands].* Nijkerk: Intro.

Swinburne, R. 1979. *The existence of God.* Oxford: Clarendon.

Tatham, G. 1951. Geography in the nineteenth century. In *Geography in the twentieth century,* edited by G. Taylor, 28-69. New York: Methuen.

Taylor, C. 1989. *Sources of the self: The making of modern identity.* Cambridge: Harvard University Press.

_____. 1992. *The ethics of authenticity.* Cambridge: Harvard University Press.

Taylor, J. F. A. 1966. *The masks of society: An inquiry into the covenants of civilization.* New York: Appleton-Century-Crofts.

Taylor, J. V. 1975. *Enough is enough.* London: SCM Press.

Thiselton, A. 1992. *New horizons in hermeneutics.* Grand Rapids: Zondervan.

Tinker, G. 1996. Spirituality, Native American personhood, sovereignty, and solidarity. In *Native and Christian: Indigenous voices on religious identity in the United States and Canada*, edited by J. Treat, 115-31. New York: Routledge.

Tuan, Y.-F. 1968. *The hydrological cycle and the wisdom of God: A theme in geoteleology*. Toronto: University of Toronto Press.

Turner, B. 1990. Outline of a theory of citizenship. *Sociology* 24, no. 2:189-217.

Van Brummelen, H. W. 1986. *Telling the next generation: Educational development in North American Calvinist Christian schools*. Lanham, Md.: University Press of America.

Van den Brink, B. 1994. Het recht van de moraal, liberalisme, en communitarisme in depolitieke filosofie [The claims of morality, liberalism, and communitarianism in depoliticized philosophy]. In *Het recht van de moraal, liberalisme, en communitarisme in de politieke filosofie [The claims of morality, liberalism, and communitarianism in political philosophy]*, edited by B. Van den Brink and W. Van Reyen, 9-23. Bussum: Coutinho.

Van der Hoeven, J. 1990. Development in the light of encounter. In *Norm and context in the social sciences*, edited by S. Griffioen and J. Verhoogt, 23-35. Lanham, Md.: University Press of America.

Van Deursen, A. 1921. Aardrijkskunde [Geography]. In *Verdieping en belijning [Deepening and defining]*. Vol. 7. Groningen: Noordhoff.

_____. 1928. Over het onderwijs in de Bijbelsche aardrijkskunde [About education in the geography of the Bible]. *Christelijk Middelbaar Onderwijs [Christian Secondary Education]* 8:405-19; 9:1-12.

_____. 1937. Aardrijkskunde [Geography]. In *Leerplan voor organisch onderwijs op de Christelijke school [Curriculum for organic education in the Christian school]*, edited by H. Dam, 77-104. Kampen: Kok.

_____. 1951. Het Christelijk beginsel en het onderwijs in de aardrijkskunde [Christian principles and education in geography]. In *Voor onderwijs en opvoeding [For education and upbringing]*. Gereformeerd School Verband, vol. 49. Kampen: Kok.

_____. 1955-56. Review of *Het karakter van de geografische totaliteit [The nature of the geographic whole]*, by G. De Jong. *Christelijk Middelbaar en Gymnasiaale Onderwijs [Christian Secondary Education]* 33:572.

Van Deursen, A., J. Overweel, and W. De Vries. 1926-39. *De Wereld en die daarin wonen: Beknopt leerboek der aardrijkskunde voor christelijke gymnasia, hoogere burgerscholen, lycea en kweekscholen* [*The world and those who live therein: Concise geography textbook for Christian high schools and teacher training colleges*]. 5 vols. Groningen: Wolters.

Van Deursen A., and W. De Vries. 1929-30. Een Christelijke leerboek der aardrijkskunde: Repliek [A Christian textbook for geography: Reply]. *Christelijk Middelbaar Onderwijs* [*Christian Secondary Education*] 10:410-18.

Van Hulst, J. W., G. Wielenga, and L. Van der Zweep. eds. 1952. *Handboek voor een Leerplan voor de Scholen met de Bijbel* [*Handbook for a curriculum for the schools with the Bible*]. Groningen: Wolters.

Van Leeuwen, M. S., ed. 1993. *After Eden: Facing the challenge of gender reconciliation*. Grand Rapids: Eerdmans.

Van Os, M., and W. J. Wieringa. eds. 1980. *Wetenschap en rekenschap, 1880-1980* [*Knowledge and explanation, 1880-1980*]. Kampen: Kok.

Van Seters, J. 1994. *The life of Moses: The Yahwist as historian in Exodus-Numbers*. Kampen: Kok-Pharos.

Vermooten, W. H., and W. Sleumer. 1929-30. Christendom en wetenschap [Christianity and science]. *Christelijk Middelbaar Onderwijs* [*Christian Secondary Education*] 10:261-66, 363-69, 403-9.

Verstraelen, F. J., A. Camps, L. A. Hoedemaker, and M. R. Spindler, eds. 1995. *Missiology: An ecumenical introduction*. Grand Rapids: Eerdmans.

Voegelin, E. 1990. *Anamnesis*. Translated and edited by G. Niemeyer. Columbia: University of Missouri Press.

Wackernagel, M., and W. Rees. 1995. *Our ecological footprint: Reducing human impact on the earth*. Gabriola Island, B.C.: New Society Publishers.

Wallace, I. 1990. *The global economic system*. London: Unwin Hyman.

Wallace, I., and D. B. Knight. 1996. Societies in space and place. In *Earthly goods: Global environmental change and social justice*, edited by F. O. Hampson and J. Reppy, 75-95. Ithaca: Cornell University Press.

Walther, Chr. 1990. Königsherrschaft Christi. In *Theologische Realencyklopädie*. Band 19, 313-23. Berlin: Walter de Gruyter.

Waltzer, M. 1994. *Thick and thin: Moral argument at home and abroad.* Notre Dame: University of Notre Dame Press.

Wang Y. L. 1935. *San zhi jing* (1223-1296). In *Chinese made easy,* translated by W. Brooks Brouner and F. Y. Mow. Leiden: Brill.

Wansink, J. M. 1928. *Aardrijkskundige bloemlezing van landen en volken (voor de Christelijke scholen)* [*Geographical anthology: Of lands and people (for the Christian schools)*]. 6 vols. Groningen: Noordhoff.

Waring, M. 1988. *If women counted: A new feminist economics.* New York: Harper & Row.

Watts, I. 1736. *The knowledge of the heavens and the earth made easy: Or, the first principles of astronomy and geography explain'd by the use of globes and maps: with a colution of the common problems by a* plain scale *and* compass *as well as by the globe.* 3d ed. London.

Weber, M. 1922. Introduction: Wirtschaftsethik der Weltreligionen. In *Gesammelte Aufsätze zur Religionssoziologie,* 237-75. Tübingen: Mohr.

_____. 1994. *Wissenschaft als Beruf 1917/1919: Politik als Beruf 1919.* Studienausgabe. Tübingen: Mohr.

Wendland, H. D. 1971. *Einführung in die Sozialethik,* Sammlung Göschen BD. 4203. Berlin: Walter de Gruyter.

Wesselius, J. W. 1995. Herodotus, vader van de bijbelse geschiedenis? [Herodotus, father of biblical history?] In *Amsterdamse Cahiers voor Exegese en Bijbelse Theology* [*Amsterdam Journal for Exegesis and Biblical Theology*]. Vol. 14, edited by K. Deurloo, 9-61. Kampen: Kok.

Westermann, C. 1974. *Genesis, Kapital 1-11: Biblischer Kommentar Altes Testaments.* Neukirchen-Vluyn: Neukirchener Verlag.

White, L. 1969. The historical roots of our ecological crisis. In *The subversive science: Essays toward and ecology of man,* edited by P. Shephard and D. McKinly, 341-51. New York: Houghton Mifflin.

Whitelam, K. W. 1996. *The invention of ancient Israel, the silencing of Palestinian history.* London: Routledge.

Wielemaker, K. 1915. 't Onderwijs in aardrijkskunde op de Christelijke school [Geographic education in the Christian school]. *Paedagogisch Tijdschrift voor het Christelijk Onderwijs* [*Pedagogical Journal for Christian Education*] 7:266-79.

_____. 1904. Review of *Op reis door Nederland: geillustreerd aardrijkskundig leesboek voor de volksschool, 1* [*On a journey through The Netherlands: Illustrated geography reader for the public school, 1*], by J. D. Bakker and F. Deelstra. *De School met de Bijbel* [*The School with the Bible*] 1, no. 39:176.

Wilkinson, L., ed. 1991. *Earthkeeping in the Nineties: Stewardship of creation.* Rev. ed. Grand Rapids: Eerdmans.

Williams, M. 1987. Sauer and "Man's role in changing the face of the earth." *Geographical Review* 77:218-31.

Williams, M., ed. 1975. *Geography and the integrated curriculum: A reader.* London: Heinemann Educational Books.

Wolters, A. M. 1985a. *Creation regained: Biblical basis for a reformational worldview.* Grand Rapids: Eerdmans.

_____. 1985b. The intellectual milieu of Herman Dooyeweerd. In *The legacy of Herman Dooyeweerd,* edited by C. T. McIntire, 1-19. Lanham, Md.: University Press of America.

_____. 1991. Gustavo Gutiérrez. In *Bringing into captivity every thought: Capita Selecta in the history of Christian evaluations of non-Christian philosophy,* edited by J. Klapwijk, 229-40. Lanham, Md.: University Press of America.

_____. 1995. Creation order: A historical look at our heritage. In *An ethos of compassion and the integrity of creation,* edited by B. J. Walsh, H. Hart, and R. E. VanderVennen, 33-48. Lanham, Md.: University Press of America.

Wolterstorff, N. P. 1984. *Reason within the bounds of religion.* 2d ed. Grand Rapids: Eerdmans.

_____. 1989. On Christian learning. In *Stained glass: Worldviews and social science,* edited by Paul A. Marshall, Sander Griffioen, and Richard J. Mouw. 56-80. Lanham, Md.: University Press of America.

_____. 1990. Response to Paul Marshall. In *Norm and context in the social sciences,* edited by S. Griffioen and J. Verhoogt, 159-63. Lanham, Md.: University Press of America.

_____. 1995. Does truth still matter? Reflections on the crisis of the postmodern university. *Crux* 31 no. 3:17-28.

Wood, A. 1994. *North-south trade, employment, and inequality: Changing fortunes in a skill-driven world.* Oxford: Clarendon.

Woods, D. 1994. Reconceptualizing economics: The contributions of Karl Polanyi. In *Political theory and Christian vision: Essays in memory of Bernard Zylstra*, edited by J. Chaplin and P. Marshall, 247-66. Lanham, Md.: University Press of America.

Worster, D. 1977. *Nature's economy: A history of ecological ideas.* Cambridge: Cambridge University Press.

Woudenberg, L. 1929-30. Esperanto. *Christelijk Middelbaar Onderwijs* [*Christian Secondary Education*] 10:286-87.

Wright, C. J. H. 1983. *Living as the people of God: The relevance of Old Testament ethics.* Leicester: Inter-Varsity Press.

Young, R. M. 1985. *Darwin's metaphor: Nature's place in Victorian culture.* Cambridge: Cambridge University Press.

Zondervan, H. 1926. Is de aardrijkskunde dood? [Is geography dead?] *Tijdschrift voor het Onderwijs in Aardrijkskunde* [*Journal for Geographic Education*] 4:155-60.

Zuidervaart, L. 1990. Response to Johan Van der Hoeven's 'Development in the light of encounter.' In *Norm and context in the social sciences*, edited by S. Griffioen and J. Verhoogt. 37-42. Lanham, Md.: University Press of America.

Zuidervaart, L., and H. Luttikhuizen. 1995. *Pledges of jubilee: Essays on the arts and culture in honor of Calvin G. Seerveld.* Grand Rapids: Eerdmans.

Index